Essentials of Inorganic Chemistry 2

D. M. P. Mingos

Sir Edward Frankland BP Professor of Inorganic Chemistry, Imperial College of Science, Technology and Medicine, London and Dean of the Royal College of Science

Series sponsor: **ZENECA**

ZENECA is a major international company active in four main areas of business: Pharmaceuticals, Agrochemicals and Seeds, Specialty Chemicals, and Biological Products.

ZENECA's skill and innovative ideas in organic chemistry and bioscience create products and services which improve the world's health, nutrition, environment, and quality of life.

ZENECA is committed to the support of education in chemistry and chemical engineering.

OXFORD NEW YORK TOKYO
OXFORD UNIVERSITY PRESS
1998

Oxford University Press, Great Clarendon Street, Oxford OX2 6DP

Oxford New York
Athens Auckland Bangkok Bogota Bombay Buenos Aires Calcutta
Cape Town Chennai Dar es Salaam Delhi Florence Hong Kong Istanbul
Karachi Kuala Lumpur Madrid Melbourne Mexico City Mumbai
Nairobi Paris São Paolo Singapore Taipei Tokyo Toronto Warsaw

and associated companies in
Berlin Ibadan

Published in the United States
by Oxford University Press Inc., New York

A catalogue record for this book is available from the British Library

Library of Congress Cataloging in Publication Data

(Data available)

ISBN 0 19 855918 6 (Pbk)

Typeset by the author

Printed in Great Britain by The Bath Press, Avon

Series Editor's Foreword

Oxford Chemistry Primers are designed to give a concise introduction to all chemistry students by providing the material that would usually form an 8–10 lecture course. As well as providing up-to-date information, this series expresses the explanations and rationales that form the framework of current understanding of inorganic chemistry.

In his first Primer (OCP28), *Essentials of Inorganic Chemistry 1*, Mike Mingos introduced an alternative style based on a glossary of key terms and concepts within the subject. This book then was a handbook of the basics of this area of chemistry. In 'Essentials 2', he covers some of the more advanced concepts that may be met in the second half of an undergraduate course. The approach again is to give rapid access to these ideas. Mike has himself contributed much to current thinking in the discipline and so is an ideal author for such a book.

John Evans
Department of Chemistry,
University of Southampton

Preface

Essentials of Inorganic Chemistry 1 which was published in 1995 attempted to provide a concise account of those concepts of inorganic chemistry which either should have been covered in a pre-university course or are introduced in the first year of a university course. *Essentials of Inorganic Chemistry* 2 adopts the same format for topics which are generally covered in the second and third years of a degree course. The coverage of more advanced topics has necessarily led to more detailed and longer individual entries. *Essentials of Inorganic Chemistry* 1 and 2 also provide the basic theoretical background for *Essential Trends in Inorganic Chemistry* which was published in 1998 and gives a very detailed account of the trends in the properties of the elements and their compounds.

I should like to thank the many colleagues who have helped me complete this book either by contributing to its production in camera ready form or reading sections of the manuscript. In particular I should like to thank John Evans who read the draft and final versions of the manuscript and who made many helpful suggestions. Bill Griffith, Paul Dyson, Brent Young, and Julian Gale read sections of the manuscript. The enormous burden of producing a camera ready version of the manuscript for OUP fell once again to Jack Barrett. I cannot thank him enough for the dedication, patience, and good humour he showed whilst completing the task. The responsibility for any inaccuracies and the subject matter remain mine alone, however. This book completes the *Essentials* trilogy and I hope not to have to bother Jack and my other colleagues for several years to come.

Imperial College of Science, Technology and Medicine D. M. P. M.
May 1998

Contents

Agostic interactions

In complexes of the transition metals which have fewer than 18 valence electrons and where the metal atom is reasonably electrophilic secondary interactions may occur between the metal and C–H, Si–H, or N–H (generally E–H) bonds associated with the ligand. These secondary interactions are generally weaker than conventional metal–ligand bonds and may result either in M···H contacts, as shown in Fig. A.1 (a), which are shorter than the sum of the Van der Waals radii or three-centre contacts in which the metal, the hydrogen atom, and the atom E participate as shown in Fig. A.1 (b). Such interactions are known as agostic.

These weak secondary bonding interactions are commonly represented by a coordinate bond either from the hydrogen atom to the metal in (a) or from the E–H bond to the metal in (b). Each agostic interaction formally increases the total valence electron count by two and may be viewed as a means of making the electron deficient molecule conform more closely to the Effective Atomic Number Rule. The interaction has approximately the same strength as a hydrogen bond.

The occurrence of agostic interactions may be demonstrated unambiguously by neutron diffraction studies (X-ray diffraction depends on the scattering by electrons and therefore is not very good for locating hydrogen atoms in crystals) or circumstantial evidence may be accumulated from spectroscopic data, e.g. for agostic C–H interactions a lowering of the (C–H) stretching frequency to approximately 2800 cm^{-1}, a shift in the ^1H n.m.r. spectrum and possibly accompanied by a reduction in the $^1J_{C-H}$ coupling constant. The neutron diffraction studies have indicated that the C–H bond is lengthened by about 3 to 10 pm from that found usually for C–H bonds (110 pm) and the metal–hydrogen distance is significantly longer than that found in metal–hydrido complexes, but sufficiently close to indicate some metal–hydrogen bonding (220–300 pm). Some typical examples of compounds which have agostic interactions are illustrated in Fig. A.2.

If the metal has filled d orbitals which are capable of donating significant amounts of electron density into the antibonding C–H bond the resulting *synergic* interactions may lead to a sufficient weakening of the C–H bond that it may be broken. In these situations an oxidative addition reaction has occurred to give a cyclo-metallated complex as exemplified by the reaction shown in Fig. A.3.

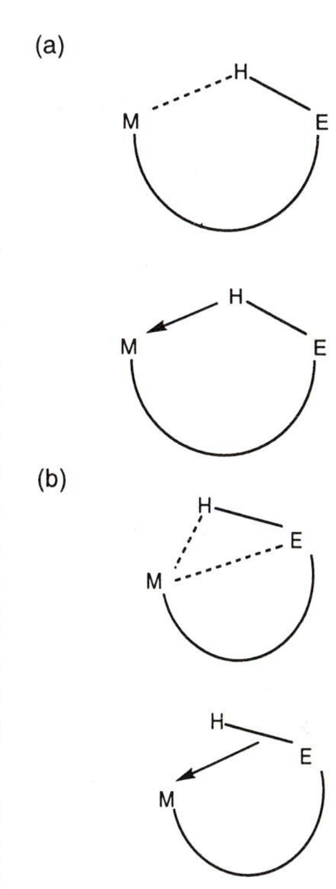

Fig. A.1 Illustrations of (a) open and (b) closed agostic interactions

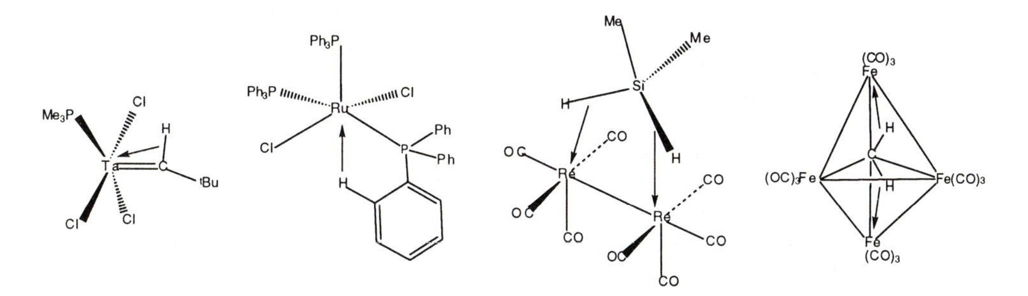

Fig. A.2 Some compounds in which agostic bonding occurs (indicated by the arrows)

See M. Brookhart and M. L. H. Green, *J. Organomet. Chem.*, (1983), **250**, 395, for a discussion of agostic bonding

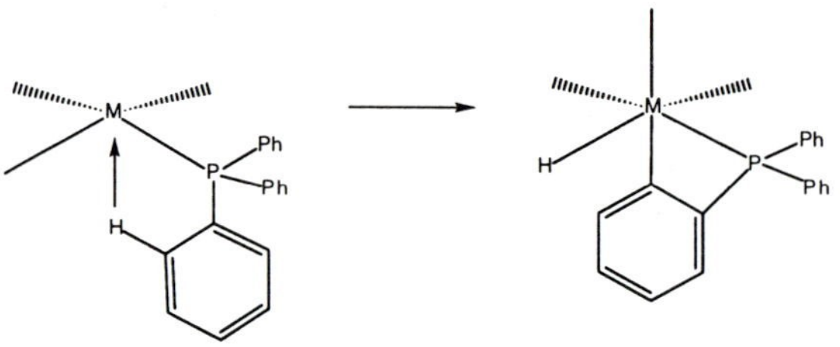

Fig. A.3 An example of the production of a cyclo-metallated complex

Angular overlap model

The angular overlap model is based on a very approximate form of molecular orbital theory and it focuses attention on the interactions between the metal *n*d valence orbitals and the ligand orbitals. It has proved to be particularly useful for accounting for the geometries of transition metal complexes and the relative positions of ligands in the spectrochemical series. The model ignores the interactions between the ligand orbitals and the metal $(n+1)$s and $(n+1)$p valence orbitals and does not incorporate effects arising from electron–electron repulsion.

For more detailed accounts of this topic see J. K. Burdett, *Molecular Shapes*, J. Wiley and Sons, New York, 1980 and A. J. Bridgeman and M. Gerloch, *Prog. Inorg. Chem.*, 1997, **45**, 179

According to second-order perturbation theory the overlap between the orbital on a ligand and the metal leads to a pair of bonding and antibonding molecular orbitals. The bonding orbital involves the in-phase overlap of the metal and ligand orbitals and the antibonding orbital the out-of-phase overlap. These molecular orbitals are stabilized/destabilized by $\pm BS^2/\Delta E$ relative to the energies of the isolated atomic orbitals as shown in Fig. A.4. S represents the overlap integral of the orbitals, ΔE represents the energy separation between the orbitals, and B is a proportionality constant. The ligand is invariably more electronegative than the metal and therefore its donor orbital lies below that of the metal on the energy scale shown in Fig. A.4.

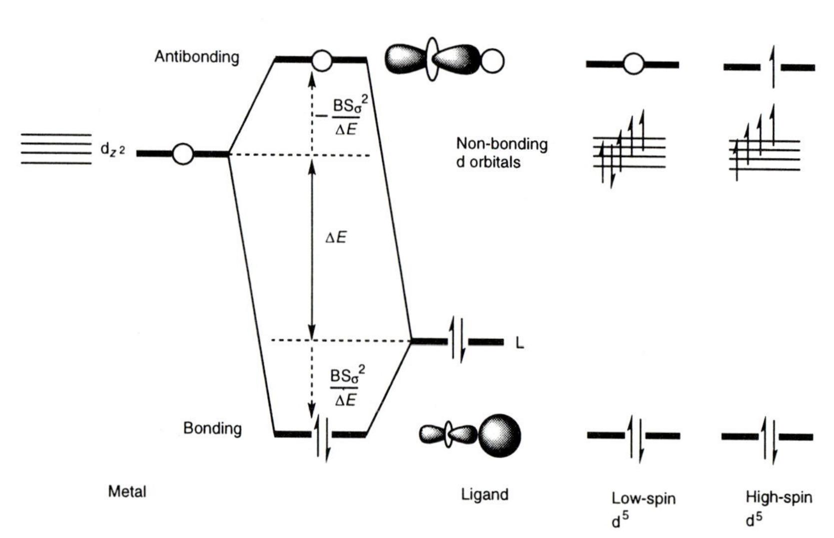

Fig. A.4 Perturbation theory analysis of bonding in M–L

The mismatch in orbital energies leads to the bonding molecular orbital being more localized on the ligand and the antibonding orbital being more localized on the metal. The extent of localization reflects the difference in electronegativities between the ligand and the metal, i.e. a larger electronegativity difference leads to more localized molecular orbitals.

For a simple complex ML (M = metal, L = ligand) where the ligand is located along the z axis the ligand overlaps exclusively with the d_{z^2} orbital and has zero overlaps with the remaining four d orbitals. The overlap integral between d_{z^2} and L is defined as S_σ (see Fig. A.5) and the orbital interactions which arise from the overlap are illustrated in Fig. A.4. The ligand orbital has overlap integrals of zero with the remaining four d orbitals because it lies on their nodal planes (see Fig. A.5). Their energies, therefore, remain unaffected and they are described as non-bonding.

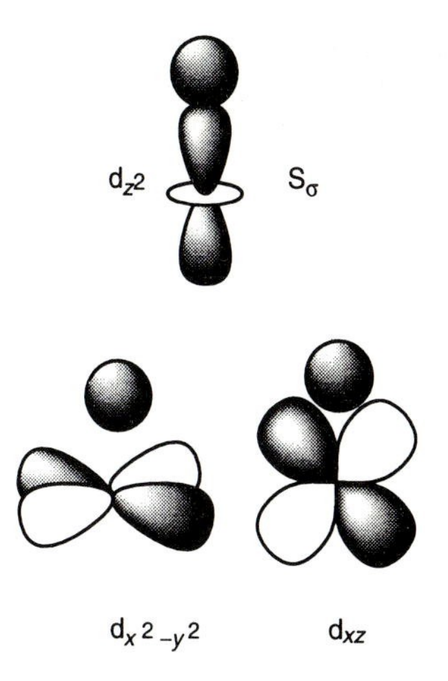

Fig. A.5 Metal–ligand overlap integrals. The upper diagram illustrates the definition of S_σ. The lower diagram shows that the overlap integrals are zero with other d orbitals

Since the ligand is a Lewis base its donor orbital is occupied by an electron pair and this electron pair is transferred to the more stable bonding orbital in the complex ML. If the metal contributes no electrons, i.e. it has a d^0 configuration, then the total stabilization energy resulting from the occupation of the bonding orbital by an electron pair is equal to $2BS_\sigma^2/\Delta E$. If the metal has $d^1 - d^8$ electron configurations then the electrons can populate the four non-bonding orbitals shown in Fig. A.4 without affecting the total stabilization energy. This assumes of course that they populate these orbitals in an *aufbau* fashion with each orbital eventually being occupied by a pair of electrons with opposing spins. Population of the antibonding orbital occurs when the metal has a d^9 configuration and the occupation of this antibonding orbital leads to a net stabilization of only $BS_\sigma^2/\Delta E$. If the antibonding orbital is doubly occupied (i.e. in a d^{10} complex) the total stabilization energy is equal to 0, since the bonding and antibonding affects cancel.

The angular overlap model ignores electron–electron repulsion effects and therefore favours an *aufbau* filling of orbitals. However, for transition metal compounds and particularly those involving first-row transition metals, the electron–electron repulsion effects are comparable to the stabilization energies arising from the covalent bonding effects described above. Therefore, 'high-spin' configurations which involve population of the antibonding d_{z^2} orbital can occur for $d^5 - d^9$ configurations. The relative stabilization energies for these 'high-spin' configurations are compared with those for 'low-spin' in Table A.1.

Table A.1

(a) Stabilization energies for 'low-spin' ML complexes

Electron configuration	Stabilization energy in units of $BS_\sigma^2/\Delta E$
d^0	2
d^1–d^8	2
d^9	1
d^{10}	0

(b) Stabilization energies for 'high-spin' ML complexes

Electron configuration	Stabilization energy in units of $BS_\sigma^2/\Delta E$
d^0	2
d^1–d^4	2
d^5	1
d^6–d^9	1
d^{10}	0

For a linear ML_2 complex with both ligands located along the z axis a similar orbital diagram results because the ligands once again overlap exclusively with d_{z^2}, but the stabilization energy associated with the bonding and antibonding orbitals is equal to $2BS_\sigma^2/\Delta E$, because each ligand causes a stabilization of $BS_\sigma^2/\Delta E$. The effects are additive and more generally in an ML_n complex the sum of the orbital stabilization energies is $nBS_\sigma^2/\Delta E$.

In a three-dimensional complex the ligands no longer overlap exclusively with d_{z^2} and a methodology must be developed for calculating the overlap integrals between a ligand in any location and the individual d orbitals. If one imagines the ligands lying on a spherical surface with the metal at the centre of the sphere then their locations are conveniently expressed in terms of the spherical polar coordinates r, θ, and ϕ as shown in Fig. A.6. If the metal–

ligand bond lengths are kept constant then only the angular variables are relevant. The overlap integral between the ligand L at θ, ϕ, and specific d orbitals is based on the trigonometric functions which define the d orbitals, i.e. the spherical harmonic functions. The relevant overlap integrals as a function of θ and ϕ are summarized in Table A.2.

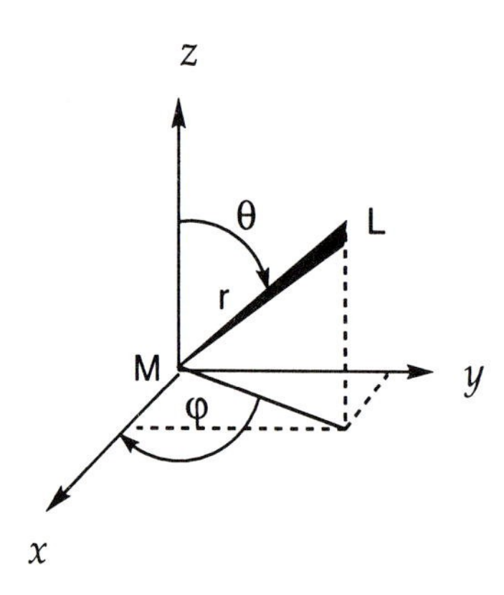

Fig. A.6 The positions of a metal atom, M, and a ligand, L, in terms of polar coordinates r, θ, and ϕ

Table A.2 Overlap integrals in terms of θ and ϕ expressed in terms of S_σ

Overlapping orbitals	Overlap integral	Overlap integral in terms of S_σ
d_{z^2}, σ	$S(d_{z^2}, \sigma)$	$[(1 + 3\cos 2\theta)/4]S_\sigma$
$d_{x^2-y^2}$, σ	$S(d_{x^2-y^2}, \sigma)$	$[(\sqrt{3/4})\cos 2\phi(1 - \cos 2\theta)]S_\sigma$
d_{xy}, σ	$S(d_{xy}, \sigma)$	$[(\sqrt{3/4})\sin 2\phi(1 - \cos 2\theta)]S_\sigma$
d_{xz}, σ	$S(d_{xz}, \sigma)$	$[(\sqrt{3/2})\cos\phi\sin 2\theta]S_\sigma$
d_{yz}, σ	$S(d_{yz}, \sigma)$	$[(\sqrt{3/2})\sin\phi\sin 2\theta]S_\sigma$

The overlap integrals are referenced relative to a ligand located at the north pole of the sphere θ = 0° and ϕ = 0°, i.e. a ligand lying along the z axis and overlapping with d_{z^2}, i.e. the substitution of θ = 0° and ϕ = 0°, into the equations for S given in Table A.2 result in the calculated overlap integrals given in the margin. This corresponds to the situation illustrated in Fig. A.5. The overlap integral for a ligand located at the south pole (θ = 0° and ϕ = 180°) is identical.

$S(d_{z^2}, \sigma)$	S_σ
$S(d_{x^2-y^2}, \sigma)$	0
$S(d_{xy}, \sigma)$	0
$S(d_{xz}, \sigma)$	0
$S(d_{yz}, \sigma)$	0

For a ligand located on the equator of the sphere θ = 90° and ϕ is variable. If the following locations are chosen ϕ = 0°, 90°, 180°, and 270°, which are appropriate (see Table A.3 for the polar coordinates of the ligands) for an octahedral complex, the overlap integrals given in Table A.4 result.

Table A.3 Polar coordinates for the ligands of an octahedral complex. The metal lies at the origin.

Ligand position	θ°	ϕ°
1	0	0
2	90	0
3	90	90
4	90	180
5	90	270
6	180	0

Table A.4 Overlap integrals for an octahedral complex

Location	A	B	C	D
ϕ (°)	0	90	180	270
$S(d_{z^2}, \sigma)$	$-1/2S_\sigma$	$-1/2S_\sigma$	$-1/2S_\sigma$	$-1/2S_\sigma$
$S(d_{x^2-y^2}, \sigma)$	$\sqrt{3}/2S_\sigma$	$-\sqrt{3}/2S_\sigma$	$\sqrt{3}/2S_\sigma$	$-\sqrt{3}/2S_\sigma$
$S(d_{xy}, \sigma)$	0	0	0	0
$S(d_{xz}, \sigma)$	0	0	0	0
$S(d_{yz}, \sigma)$	0	0	0	0

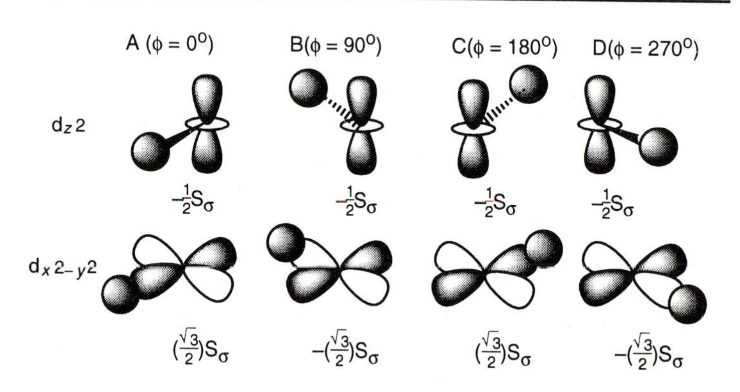

Fig. A.7 Positions of ligand orbitals and the d_{z^2} orbital of a metal atom and the values of the overlap integral

The illustrations shown in Fig. A.7 reinforce in a pictorial manner the results of the calculations. A ligand located on the equator overlaps with the 'collar' of the d_{z^2} orbital and consequently the sign of the overlap integral and its magnitude changes relative to that calculated for a polar position. Since the d_{z^2} orbital is rotationally symmetric about the z axis the overlap integrals are independent of ϕ and equal to $-\frac{1}{2}S_\sigma$. The $d_{x^2-y^2}$ orbital has two nodal planes which intersect on the z axis and therefore the overlap of the orbitals lying on the equator are dependent on ϕ. Specifically the lobes of the orbital have their maxima at $\phi = 0°$, $90°$, $180°$, and $270°$ and therefore at these locations the overlap integrals are equal and have a magnitude of $\frac{\sqrt{3}}{2}S_\sigma$, but the signs change every 90° because of the presence of the nodal planes at $\phi =$ 45°, 135°, 225°, and 315°, i.e. for these angles the calculated overlap integrals are equal to zero. The d_{xy}, d_{xz}, and d_{yz} orbitals have zero overlaps with the ligands in an octahedral complex (see Table A.4).

Therefore, the mathematical expressions given in Table A.2 define the positions of the nodal planes and the shapes of the lobes of the orbitals and thereby provide a trigonometric means of calculating how the overlap integrals between a ligand and the metal d orbitals vary as the ligand migrates across the surface of the imaginary ligand sphere.

The relevant orbital interaction energies for an octahedral complex may be calculated from the overlap expressions given in Table A.4. The results are summarized in Fig. A.8 and form the basis for the detailed molecular orbital diagram shown in Fig. A.9.

The d_{z^2} and $d_{x^2-y^2}$ orbitals are destabilized by $3BS_\sigma^2/\Delta E$ relative to the d_{xy}, d_{xz}, and d_{yz} orbitals, which remain non-bonding. It is noteworthy that the degeneracy of the d_{z^2} and $d_{x^2-y^2}$ orbitals is retained since they interact equally with the ligand orbitals in an octahedral complex. In group theoretical terms they are described as a pair of orbitals with e_g symmetry. The non-bonding d_{xy}, d_{xz}, and d_{yz} orbitals correspond to a triply degenerate set with t_{2g} symmetry. The ligand linear combinations which match the d_{z^2} and $d_{x^2-y^2}$ orbitals are also stabilized by $3BS_\sigma^2/\Delta E$.

Fig. A.8 Total stabilization energies for the bonding molecular orbitals resulting from the overlap of the d_{z^2} and $d_{x^2-y^2}$ orbitals with the ligand orbitals in an octahedral complex. The antibonding combinations are localized on the metal and are destabilized by $3BS_\sigma^2/\Delta E$

Table A.5
(a) Stabilization energies (in units of $BS_\sigma^2/\Delta E$) for 'low-spin' ML_6 octahedral complexes

Electron configuration	Stabilization energy
d^0	12
d^1–d^6	12
d^7	9
d^8	6
d^9	3
d^{10}	0

(b) Stabilization energies (in units of $BS_\sigma^2/\Delta E$) for 'high-spin' ML_6 octahedral complexes

Electron configuration	Stabilization energy
d^0	12
d^1– d^3	12
d^4	9
d^5	6
d^6– d^8	6
d^9	3
d^{10}	0

Table A.6 Pairing energies for some ions /kJ mol^{-1}

Cr^{2+}	281
Mn^{2+}	305
Fe^{2+}	211
Co^{2+}	269
Fe^{3+}	359
Co^{3+}	251

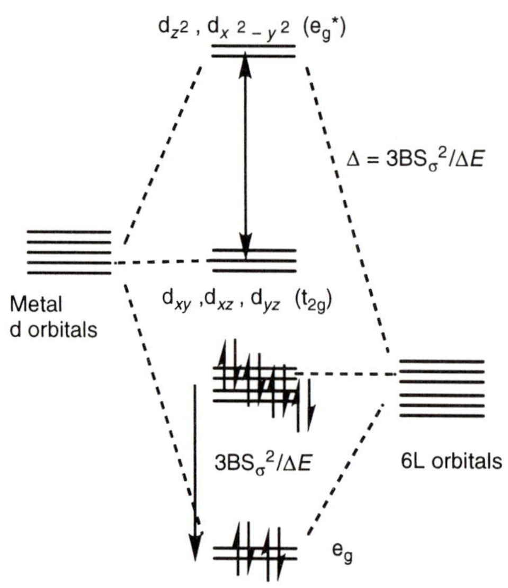

Fig. A.9 Molecular orbital diagram for an octahedral complex, showing the locations of the six pairs of ligand electrons

The occupation of the six molecular orbitals localized primarily on the ligands by the electrons which were initially defined as lone pair electrons results in a total stabilization energy of $12BS_\sigma^2/\Delta E$, i.e. $2nBS_\sigma^2/\Delta E$ for a d^0 complex. For metal ions with d^1 – d^{10} electron configurations the orbitals localized mainly on the metal, i.e. t_{2g} and e_g^*, are occupied. The similar energies associated with the covalent bonding between the metal d orbitals and the ligands and electron–electron repulsion energies once again gives rise to high and low spin possibilities. The relative total stabilization energies are summarized in Table A.5. The occurrence of a high or low spin configuration in a specific complex depends on the magnitudes of the d orbital splitting energy, $3BS_\sigma^2/\Delta E$, relative to the pairing energy required to move an electron from an e_g^* orbital into a t_{2g} orbital.

Some typical pairing energies and orbital splitting energies for aqua-complexes are summarized in Tables A.6 and A.7. It is apparent that for complexes of water the pairing energies are greater than the ligand field splitting energies. Therefore, aqua complexes of the first row transition metals are rarely observed as low-spin complexes. For ligands which cause splittings which are greater than the pairing energies low spin complexes are observed. The CN^- ligand which causes particularly large splittings always forms low spin complexes (see Table A.7).

The d orbital splitting energies in a series of complexes may be established experimentally by measuring the energy change required to promote an electron from the t_{2g} to the e_g^* orbitals.

The d orbital splitting energies occur approximately in the visible region of the electromagnetic spectrum and therefore they may be obtained from electronic spectral data. The electronic transitions involve the promotion of electrons within the d sub-shell and are therefore forbidden by the Laporte selection rule, i.e. for an electronic transition in an atom to be allowed the value of $\Delta l = \pm 1$, i.e. s → p, p → d, d → f, etc. This selection rule carries over to metal complexes as long as they retain a centre of symmetry, i.e. a centre of inversion retains the distinctions between s and p, p and d orbitals, etc. The second important selection rule is the spin selection rule which states that $\Delta s = 0$. In general this selection rule can be adhered to in the d–d electronic transitions for all transition metal ions except high spin d^5.

Weak d–d transitions are observed for the great majority of transition metal complexes in the visible region, because although the transitions are electric dipole forbidden the vibrations associated with the metal–ligand bonds can destroy the centre of symmetry in the octahedral molecules at the instant of the transition and consequently the transitions are not completely forbidden.

If the ligand is varied for a given metal ion the resulting series is described as the *spectrochemical series*. This series is found to be similar for all transition metal ions and in a truncated form it is summarized below:

$I^- < Br^- < S^{2-} < \underline{S}CN^- < Cl^- < NO_3^- < F^- < C_2O_4^{2-} < H_2O < \underline{N}CS^- < NH_3 < en < bipy < phen < NO_2^- < PPh_3 < \underline{C}O < \underline{C}N^-$

This series cannot be interpreted simply in terms of electrostatic effects, i.e. the crystal field model, since neutral ligands such as CO and NH_3 appear higher in the spectrochemical series than small anionic ligands such as F^-. According to the angular overlap model the strength of covalent bonding in such complexes depends primarily on the *electronegativity* of the donor atom (the lower the electronegativity the more efficient the electron pair donation in the coordinate M–L bond), and the π-bonding character of the ligand.

The d orbital splitting energy increases as the electronegativity of the donor atom is reduced, i.e. as the energies of the metal and ligand orbitals become closer in energy terms ΔE decreases and the splitting energy, $BS_\sigma^2/\Delta E$, becomes larger.

Electronegativity trends are: F > O > N > C and

Cl > S > P > Si

leading to an increase in $BS_\sigma^2/\Delta E$ as the electronegativity decreases. The increase in the formal oxidation state of the metal effectively increases the electronegativity of the metal and thereby decreases the metal–ligand electronegativity difference. This also has the effect of increasing the splitting energy.

Although the metal–ligand σ–covalent bond occurs through the metal d_{z^2} and $d_{x^2-y^2}$ orbitals, the π-bonding ability of the ligand is important because the metal d_{xz}, d_{yz}, and d_{xy} orbitals have the correct symmetry properties to overlap with the π-orbitals of the ligands as shown in Fig. A.10. The standard π-overlap integral, S_π, is defined in an analogous way to that described previously for S_σ (see Fig. A.10). The stabilization resulting from the overlap between the ligand π orbitals and one of the t_{2g} orbitals results in a total stabilization energy of $4B_\pi S_\pi^2/\Delta E$, where B_π represents the proportionality constant.

The effect of overlap between the ligand π-orbitals and the metal d_{xz}, d_{yz}, and d_{xy} orbitals depends on whether the ligands are π-donors or π-acceptors.

Table A.7 Orbital splitting energies /kJ mol^{-1}

$[V(OH_2)_6]^{2+}$	141
$[Cr(OH_2)_6]^{2+}$	167
$[Mn(OH_2)_6]^{2+}$	93
$[Fe(OH_2)_6]^{2+}$	124
$[Co(OH_2)_6]^{2+}$	111
$[V(OH_2)_6]^{3+}$	215
$[Cr(OH_2)_6]^{3+}$	214
$[Mn(OH_2)_6]^{3+}$	251
$[Fe(OH_2)_6]^{3+}$	164
$[Co(OH_2)_6]^{3+}$	218
$[Ti(CN)_6]^{3-}$	280
$[V(CN)_6]^{3-}$	318
$[Cr(CN)_6]^{3-}$	419
$[Fe(CN)_6]^{3-}$	416
$[Co(CN)_6]^{3-}$	404

(a)

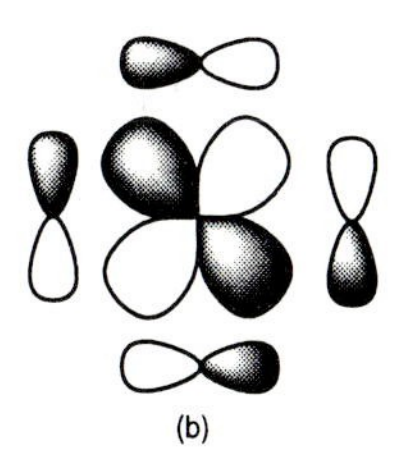
(b)

Fig. A.10 (a) Definition of the π-overlap integral between a t_{2g} metal orbital and a ligand p orbital, S_π, and (b) the bonding interaction between a t_{2g} metal orbital and four ligand p orbitals

π-donors have filled orbitals which lie below the d_{xz}, d_{yz}, and d_{xy} orbitals and consequently have the effect of reducing the splitting energy, whereas π-acceptors have empty orbitals which lie above the d_{xz}, d_{yz}, and d_{xy} orbitals and have the effect of increasing the splitting energy. The orbital interactions responsible for these effects are illustrated in Fig. A.11 and typical π-donors and acceptors are summarized in the margin.

π-donors
F^-, O^{2-}, N^{3-}
Cl^-, S^{2-}

no π-character
NH_3, H^-, en

π-acceptors
CO, CN^-, NO^+
phen, bipy

Ligands which are high in the spectrochemical series are either good σ-donors and π-acceptors or poor σ-donors but very good π-acceptors (e.g. CO). Those which are low in the spectrochemical series are poor σ-donors and good π-donors.

The splitting energy is also increased by better metal–ligand overlap and this effect is most prominent as the metal atom is changed. Specifically the splitting energy increases in the order 3d < 4d < 5d because the increase in the principal quantum number leads to enlarged metal d orbitals which are able to overlap more effectively with the ligand orbitals, i.e. S_σ and S_π increase in the $BS^2/\Delta E$ terms.

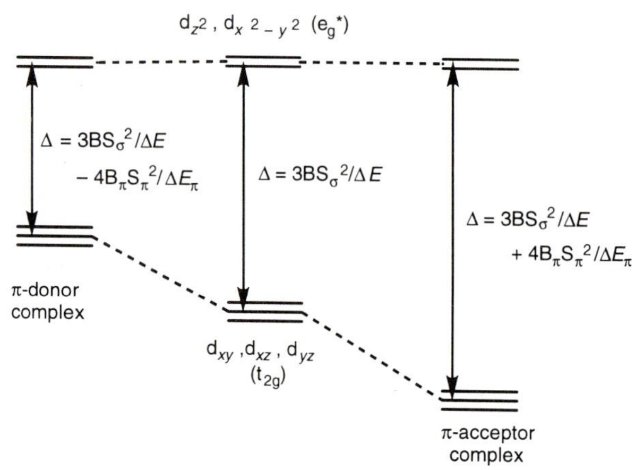

Fig. A.11 Effects of π-bonding on Δ in a π-donor and a π-acceptor complex

The following trends in Δ are therefore observed:

$$Fe^{3+} < Co^{3+} < Rh^{3+} < Ir^{3+} < Pt^{4+}$$

The increased covalency in the metal–ligand bonds also leads to an increase in Δ across the transition series:

$$Mn^{2+} < V^{2+} < Co^{2+} < Fe^{2+} < Ni^{2+}$$

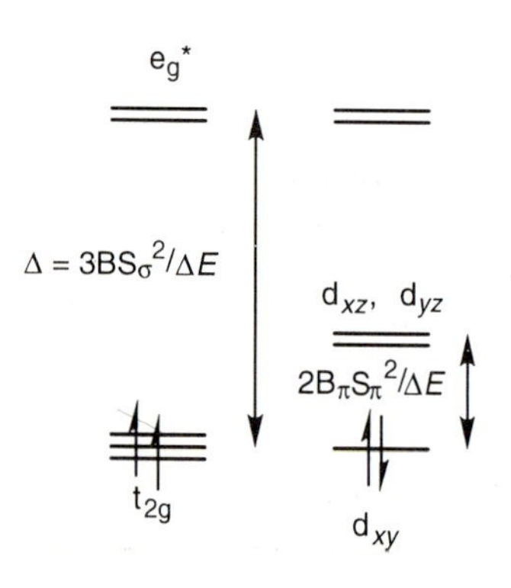

Fig. A.12 The orbital splitting diagram for *trans*-$[MO_2(OH_2)_4]$ showing the effect of π-bonding

The angular overlap model may be used to evaluate the effect of having one or two π-donor or acceptors in an octahedral complex. For example, in *trans*-$[MO_2(OH_2)_4]$ and $[MO(OH_2)_5]$, the p_π orbitals of the oxo-ligands overlap exclusively with d_{xz} and d_{yz}, and d_{xy} remains non-bonding. The relevant orbital splitting diagram for $[MO_2(OH_2)_4]$ is shown in Fig. A.12 and assumes that S_σ for the oxo- and aqua-ligands are equal. Clearly, if $B_\pi S_\pi$

is large, i.e. the oxo-ligand is a good π-donor, the π-bonding effects introduce a secondary gap in the spectrum of molecular orbitals within the t_{2g} set.

If the gap is larger than the pairing energy a low spin configuration is favoured. *Trans*-$[MO_2(OH_2)_4]$ complexes with d^2 configurations are common and are always diamagnetic. A similar analysis may be applied to d^2 mono-oxo-complexes which are also diamagnetic.

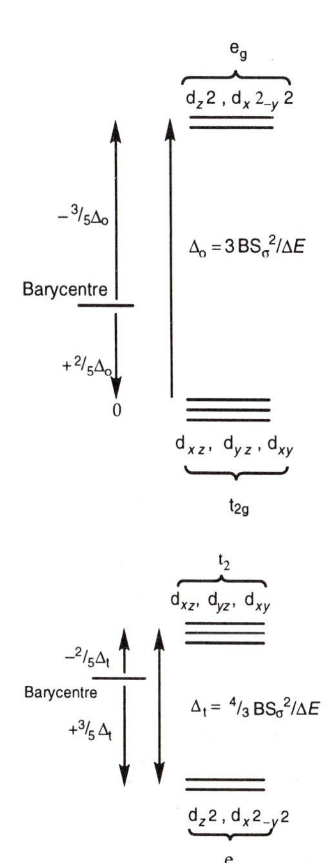

Fig. A.13 Octahedral and tetrahedral splittings of the d orbitals. Note that the ratio of the orbital splittings is equal to $\frac{9}{4}$

Consequences of d-orbital splittings in octahedral complexes

The interpretation of the spectrochemical series requires an appreciation of the strength of the covalent interactions in complexes. However, there are many properties associated with transition metal complexes which require only the knowledge that there is a splitting of the d orbitals and an empirical knowledge of the spectrochemical series.

The molecular orbital diagram illustrated in Fig. A.13 has its energy scale referenced to the non-bonding d orbitals. However, this is arbitrary and it is at times more convenient to define the zero point as the weighted mean of the energies of all of the d orbitals.

The orbitals shown in Fig. A.13 correspond to those orbitals which have a high proportion of d orbital character. The energy difference of $3BS_\sigma^2/\Delta E$ between these orbital sets may be represented by Δ. For a weighted zero point on the energy scale (known as a barycentre) the relative energies of the d orbitals are $+\frac{2}{5}\Delta(t_{2g})$ and $-\frac{3}{5}\Delta(e_g)$ as shown in Fig A.13. An identical orbital splitting diagram may be derived from the electrostatic *crystal field theory*. This results because both models incorporate the basic symmetry aspects of the problem.

Magnetic properties

Since high spin and low spin configurations have different total numbers of unpaired spins (for example, the low spin configuration $(t_{2g})^4$ is associated with a total spin $S = 1$; but the high spin configuration $(t_{2g})^3(e_g)^1$ gives a total spin $S = 2$) they may be distinguished by their paramagnetic moments, which are proportional to $\sqrt{(4S(S+1))} = \sqrt{[n(n+2)]}$ if orbital angular momentum effects are ignored. Examples of some octahedral high spin and low spin complexes are given in Table A.8.

Table A.9 summarizes the magnetic moments of octahedral transition metal complexes with d^1–d^9 electronic configurations. The examples illustrate clearly the alternative high spin (hs) and low spin (ls) possibilities for d^4, d^5, d^6, and d^7 configurations. It is noteworthy that for these examples the high spin variant is associated with ligands that are low in the spectrochemical series, such as F^-, and the low spin with ligands such as CN^-. Table A.9 also suggests that the spin-only formula which neglects orbital contributions to the magnetic moment works very well for the majority of the examples. Indeed, the variations observed for d^4 (ls), d^5 (ls), d^6 (hs), d^7 (hs), and d^8 may be related to the orbital contribution which has been neglected here.

Table A.8 Examples of some octahedral high and low spin complexes

	High spin	Low spin
d^4	$t_{2g}^3e_g^1$	t_{2g}^4
	$[MnF_6]^{3-}$	$[Mn(CN)_6]^{3-}$
d^5	$t_{2g}^3e_g^2$	t_{2g}^5
	$[Fe(C_2O_4)_3]^{3-}$	$[Fe(CN)_6]^{3-}$
d^6	$t_{2g}^4e_g^2$	t_{2g}^6
	$[Fe(OH_2)_6]^{2+}$	$[Fe(CN)_6]^{4-}$

Thermodynamic effects

Table A.10 summarizes the stabilization energies of high spin d^0–d^{10} octahedral complexes. The effects rize from d^0 and d^5 to a maximum at d^3 and d^8. These differences in stabilization energies as a function of the d electron configuration are reflected in a range of thermodynamic data.

Table A.9 Magnetic properties of some octahedral complexes (n = number of unpaired spins)

d configuration		$S = \frac{1}{2}n$	$\mu_{eff} = 2\sqrt{S(S+1)}$	Example	μ_{eff} observed
d^1	$(t_{2g})^1$	$\frac{1}{2}$	1.73	K_3TiF_6	1.70
d^2	$(t_{2g})^2$	1	2.83	K_3VF_6	2.79
d^3	$(t_{2g})^3$	$\frac{3}{2}$	3.87	$[Cr(NH_3)_6]Cl_3$	3.85
d^4 hs	$(t_{2g})^3(e_g)^1$	2	4.90	K_3MnF_6	4.95
d^4 ls	$(t_{2g})^4$	1	2.83	$K_3Mn(CN)_6$	3.2
d^5 hs	$(t_{2g})^3(e_g)^2$	$\frac{5}{2}$	5.92	Na_3FeF_6	5.85
d^5 ls	$(t_{2g})^5$	$\frac{1}{2}$	1.73	$K_3Fe(CN)_6$	2.4
d^6 hs	$(t_{2g})^4(e_g)^2$	2	4.90	K_3CoF_6	5.53
d^6 ls	$(t_{2g})^6$	0	0	$K_4Fe(CN)_6$	0
d^7 hs	$(t_{2g})^5(e_g)^2$	$\frac{3}{2}$	3.87	CoI_2	5.03
d^7 ls	$(t_{2g})^6(e_g)^1$	$\frac{1}{2}$	1.73	$[Co(diars)_3](ClO_4)_2$	1.92
d^8	$(t_{2g})^6(e_g)^2$	1	2.83	$[Ni(NH_3)_6]Cl_2$	2.8
d^9	$(t_{2g})^6(e_g)^3$	$\frac{1}{2}$	1.73	$[Cu(OH_2)_6]^{2+}$	1.9

Table A.10 Stabilization energies for high spin octahedral complexes

d configuration	Stabilization energy
d^0	0
d^1, d^6	$\frac{2}{5}\Delta$
d^2, d^7	$\frac{4}{5}\Delta$
d^3, d^8	$\frac{6}{5}\Delta$
d^4, d^9	$\frac{3}{5}\Delta$
d^5, d^{10}	0

The hydration enthalpies for M^{2+} and M^{3+} ions and the enthalpies of formation of the halides MX_2 and MX_3 are expected to become more negative across the series because of the contraction in radius which results from the increasing effective nuclear charge.

This is indeed observed as shown in Figs A.14 and A.15, but superimposed on the anticipated smooth curves are perturbations which give the curves a 'double humped' appearance. The additional contribution which results in this appearance can be related to the stabilization energies given in Table A.10. If these relevant stabilization energies are subtracted the resulting points more closely fit a smooth curve which passes through the points for d^0, d^5, and d^{10} configurations for which the stabilization energies are zero (see Fig. A.14 for example).

$AsMe_2$

$AsMe_2$

Diars

Covalent radii

Table A.11 presents the covalent radii of the transition metals in their II and III formal oxidation states. The overall contraction in covalent radii across the series results from the increase in effective nuclear charge, but superimposed on it are the orbital splitting effects. The e_g orbitals are antibonding and consequently their occupation leads to longer metal–ligand bonds. Therefore, the radii rise for d^4 and d^5 high spin complexes and d^9 and d^{10} complexes where the e_g orbitals are populated.

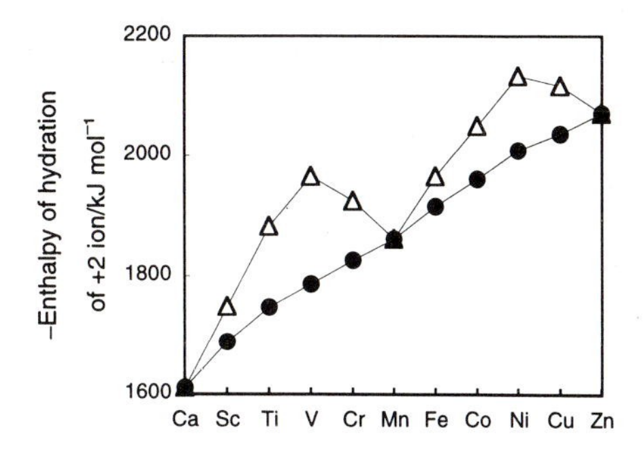

Fig. A.14 Hydration enthalpies (open triangles are observed values, filled circles are values for which the d orbital stabilization energies have been subtracted) for M^{2+} ions of the first transition series

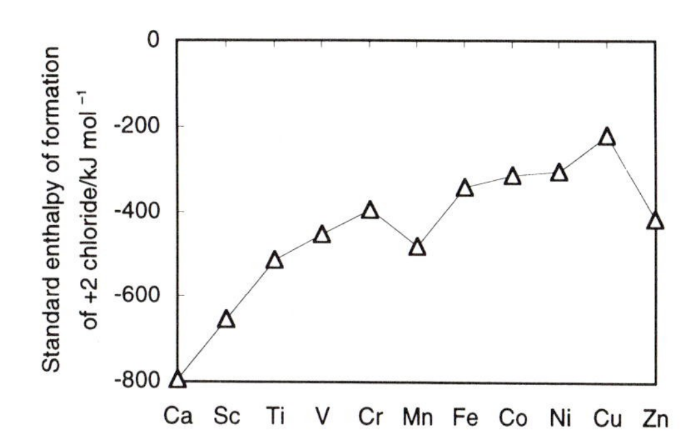

Fig. A.15 Enthalpies of formation for MCl_2 of the first transition series

Table A.11 The M^{II} and M^{III} covalent radii for some first row transition elements

Metal	Covalent radii/pm and d electron configurations	
	M^{II}	M^{III}
Ti	166 $(t_{2g})^2$	147 $(t_{2g})^1$
V	159 $(t_{2g})^3$	144 $(t_{2g})^2$
Cr	160 $(t_{2g})^3(e_g)^1$	142 $(t_{2g})^3$
Mn	163 $(t_{2g})^3(e_g)^2$	145 $((t_{2g})^3(e_g)^1$
Fe	158 $(t_{2g})^4(e_g)^2$	145 $(t_{2g})^3(e_g)^2$
Co	155 $(t_{2g})^5(e_g)^2$	141 $(t_{2g})^4(e_g)^2$
Ni	149 $(t_{2g})^6(e_g)^2$	140 $(t_{2g})^5(e_g)^2$
Cu	153 $(t_{2g})^6(e_g)^3$	

The different characteristics of the t_{2g} and e_g orbitals also leads to dramatic differences in radii between high and low spin complexes. Some typical values are summarized in Table A.12. It is noteworthy that the low spin complexes always have the smaller radii, because the depopulation of the e_g antibonding orbitals results in strengthening of the metal–ligand bonds.

Table A.12 Covalent radii of high and low spin configurations of d^{4-7} metal ions

d configuration	High spin	Radius/pm	Low spin	Radius/pm
d^4 Cr^{2+}	$(t_{2g})^3(e_g)^1$	160	$(t_{2g})^4$	150
d^5 Mn^{2+}	$(t_{2g})^3(e_g)^2$	163	$(t_{2g})^5$	147
d^6 Fe^{2+}	$(t_{2g})^4(e_g)^2$	158	$(t_{2g})^6$	141
d^7 Co^{2+}	$(t_{2g})^5(e_g)^2$	155	$(t_{2g})^6(e_g)^1$	145

Geometric consequences

Since the angular overlap model provides a convenient methodology for calculating the d orbital splittings in transition metal complexes it is very useful for estimating the relative stabilities of alternative coordination geometries and tracing the effect on the d orbital manifold of removing or adding ligands. The pairwise nature of the stabilization energy involving a ligand and a metal d orbital makes it particularly easy to evaluate the effect of removing ligands from a coordination sphere.

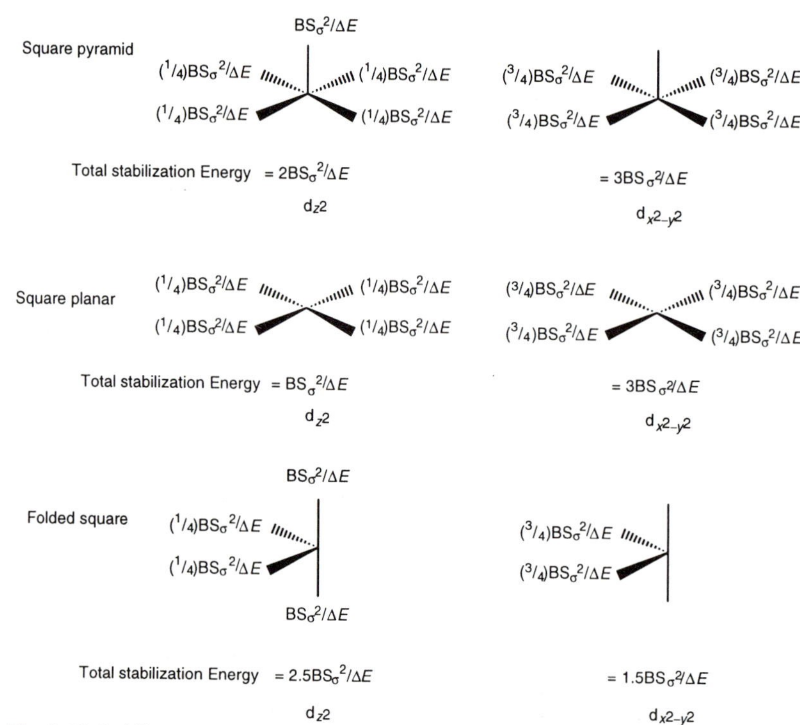

Fig. A.16 Stabilization energies for the bonding molecular orbitals resulting from the d_{z^2} and $d_{x^2-y^2}$ orbitals in square pyramidal, square planar, and folded square environments

For example, Fig. A.16 traces the effect of successively removing ligands from an octahedron. The loss of a single ligand along the z axis, which generates a square pyramid, results in the loss of $BS_\sigma^2/\Delta E$ stabilization energy for the d_{z^2} orbital, but has no effect on $d_{x^2-y^2}$ because there is zero overlap between a ligand located on the z axis and $d_{x^2-y^2}$.

Fig. A.16 also indicates that the removal of a second ligand from the z axis, resulting in a square planar geometry, causes a further decrease of $BS_\sigma^2/\Delta E$ in the stabilization energy associated with d_{z^2}, but does not influence $d_{x^2-y^2}$. However, if two *cis*-ligands are removed from the octahedron to generate a four coordinate folded square geometry then the loss of stabilization for d_{z^2} is only $0.5BS_\sigma^2/\Delta E$, because the ligands only interact with the 'collar' of the d_{z^2} orbital, whereas the loss for $d_{x^2-y^2}$ is $1.5BS_\sigma^2/\Delta E$.

These stabilization energies may be used to construct the energy level diagram illustrated in Fig. A.17. This figure traces the effect of ligand loss on the energies of the d orbitals in the d manifold and also the total stabilization energies associated with the metal ligand bonding molecular orbitals. Particularly noteworthy is the loss of $2BS_\sigma^2/\Delta E$ for the latter each time that a ligand is taken away from the coordination sphere. In Fig. A.17 $B/\Delta E$ has been replaced by β because if it is assumed that all the complexes in a series have the same ligands and equal bond lengths there is no variation in the proportionality constant and the energy difference, ΔE, between metal and ligand.

Short cuts

The d orbital splittings for these octahedrally based fragments may be quickly derived by recalling the following key stabilization energy contributions:

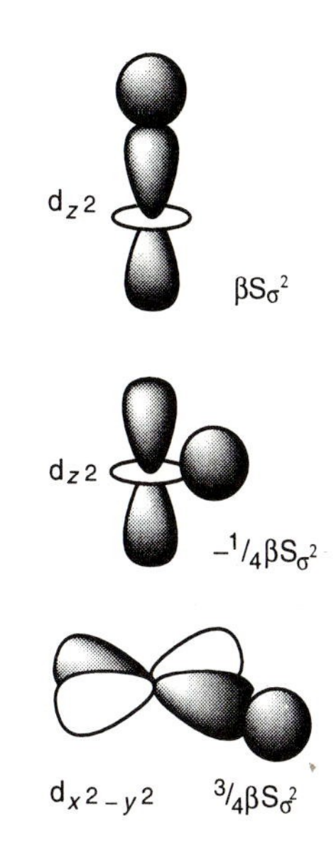

The stabilization energy associated with a specific d orbital is the sum of individual contributions from each metal–ligand bond.

The final d orbital splitting energies may be checked by recognizing that in a complex ML_n the total sum of the d orbital energies must equal $-n\beta S_\sigma^2$

The folded square geometry is also described as a saw-horse geometry

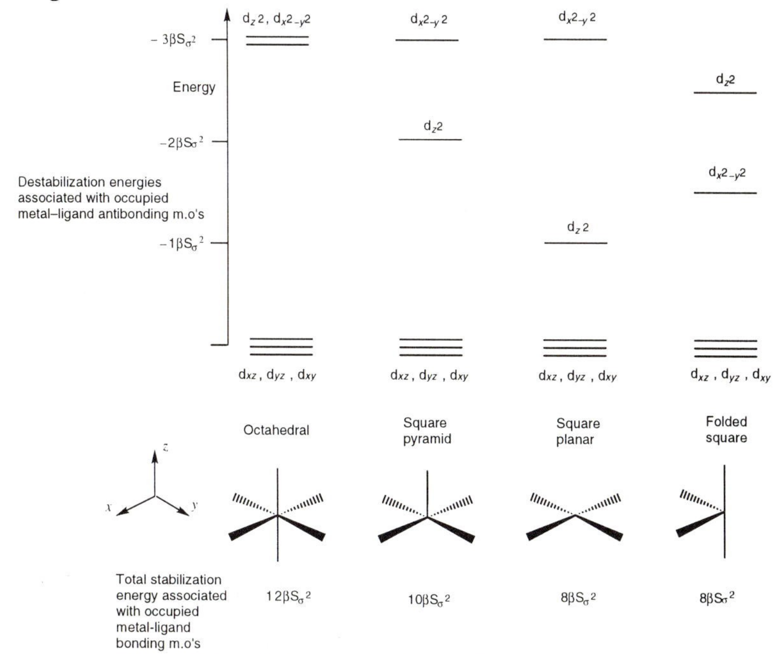

Fig. A.17 Orbital splittings in octahedrally related complexes

> The total stabilization energy associated with the doubly occupied metal–ligand bonding molecular orbitals in a complex ML_n is always equal to $2nBS_\sigma^2/\Delta E$ ($= 2n\beta S_\sigma^2$).

The way in which the d orbital splitting energies may be constructed for a series of complexes by the stepwise removal of ligands may be extended to other series of molecules. For example, for a trigonal bipyramid the relevant polar coordinates and overlap integrals calculated from Table A.2 for the equatorial ligands are given in Tables A.13 and A.14. The resulting stabilization energies and d orbital splitting diagrams are shown in Figs. A.18 and A.19.

Table A.13 Polar coordinates (°) for a trigonal bipyramidal complex

Ligand position	θ	ϕ
1	0	0
2	90	0
3	90	120
4	90	240
5	180	0

Table A.14 Calculated overlap integrals for the equatorial ligands in a trigonal bipyramid

Location	2	3	4
ϕ (°)	0	120	240
$S(d_{z^2}, \sigma)$	$-1/2S_\sigma$	$-1/2S_\sigma$	$-1/2S_\sigma$
$S(d_{x^2-y^2}, \sigma)$	$\sqrt{3}/4S_\sigma$	$-\sqrt{3}/4S_\sigma$	$\sqrt{3}/4S_\sigma$
$S(d_{xy}, \sigma)$	0	$3/4S_\sigma$	$3/4S_\sigma$
$S(d_{xz}, \sigma)$	0	0	0
$S(d_{yz}, \sigma)$	0	0	0

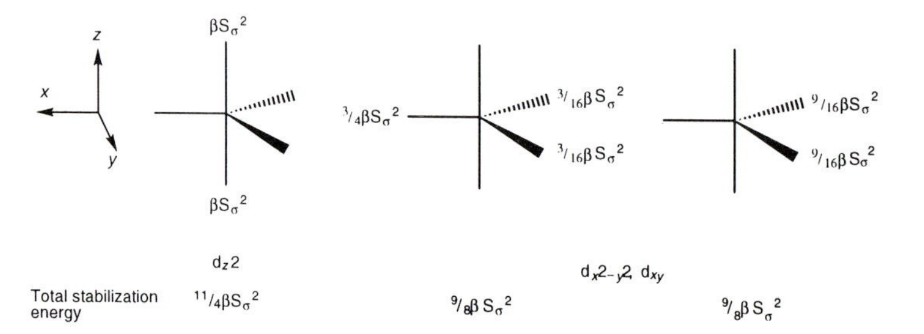

Fig. A.18 Total stabilization energies for the d orbitals participating in the equatorial plane of a trigonal bipyramidal complex

Table A.15 Polar coordinates (°) for a tetrahedral complex

Ligand position	θ	ϕ
1	109.47/2	45
2	109.47/2	225
3	180–109.47/2	135
4	180–109.47/2	315

Table A.16 Overlap integrals for a tetrahedral complex

Location	1	2	3	4
ϕ (°)	45	225	135	315
$S(d_{z^2}, \sigma)$	0	$0\sqrt{3}$	0	0
$S(d_{x^2-y^2}, \sigma)$	0	0	0	0
$S(d_{xy}, \sigma)$	$1/\sqrt{3}S_\sigma$	$1/\sqrt{3}S_\sigma$	$-1/\sqrt{3}S_\sigma$	$-1/\sqrt{3}S_\sigma$
$S(d_{xz}, \sigma)$	$1/\sqrt{3}S_\sigma$	$-1/\sqrt{3}S_\sigma$	$1/\sqrt{3}S_\sigma$	$-1/\sqrt{3}S_\sigma$
$S(d_{yz}, \sigma)$	$1/\sqrt{3}S_\sigma$	$-1/\sqrt{3}S_\sigma$	$-1/\sqrt{3}S_\sigma$	$1/\sqrt{3}S_\sigma$

For a tetrahedral complex the polar coordinates for the ligands are given in Table A.15 and the overlap integrals calculated for the trigonometric expressions in Table A.2 are summarized in Table A.16. The overlap between a ligand orbital and a metal d_{xy} orbital for a tetrahedral complex is shown in Fig. A.20. The relevant interaction diagrams and overlaps are illustrated in Fig. A.21. In a tetrahedral complex the orbital energies of the d_{xz}, d_{yz}, and d_{xy} and the $d_{x^2-y^2}$ and d_{z^2} orbitals are reversed relative to those of the octahedron and the orbital splitting is $\frac{4}{9}\,\Delta_{\text{octahedral}}$.

The orbital splitting diagrams calculated above may be used to provide some insight into the geometric preferences for particular coordination numbers. Table A.17 summarizes the relevant data for four coordinate complexes with tetrahedral, square planar, folded square and trigonal pyramidal geometries. The electron occupations of the d orbitals are indicated in the left hand column of the Table. d^0 (00000), d^5 (high spin, 11111), and d^{10} (22222) complexes have equal stabilization energies and therefore their preferred geometries are not influenced by the relative splittings of the d orbitals. The stabilization energies are $8\beta S_\sigma^2$, $4\beta S_\sigma^2$, and $0\beta S_\sigma^2$ respectively. The maximum stabilization energies are obtained when the d orbitals are unoccupied and the maximum stabilization associated with the metal-ligand molecular orbitals is achieved ($2n\beta S_\sigma^2$ with $n = 4$ in this instance). The half filled shell gives rise to only $n\beta S_\sigma^2$ because the antibonding orbitals localized on the metal are half filled and the d^{10} have no net stabilization energies because these orbitals are doubly occupied.

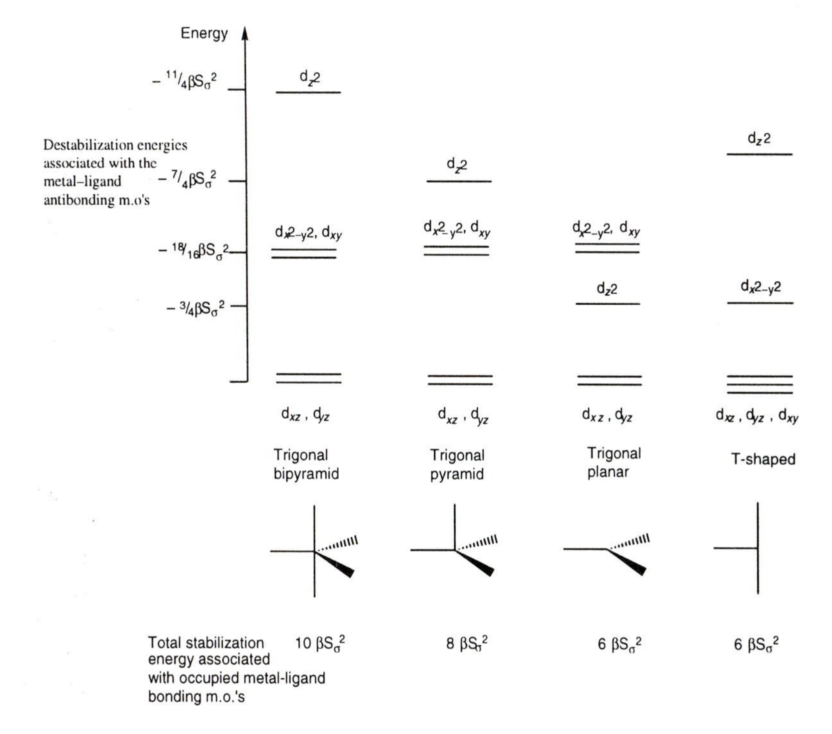

Fig. A.19 Orbital splittings for complexes related to the trigonal bipyramid

Table A.17 A summary of net stabilization energies (in units of βS_σ^2) for four coordinate complexes

d electron configuration[a]		Tetrahedron	Square planar	Folded square	Trigonal pyramid
d^{10}	22222	0	0	0	0
d^9	22221	1.33	**3.00**	2.50	1.75
d^8	22220	2.67	**6.00**	5.00	3.50
d^8	22211	2.67	**4.00**	**4.00**	2.875
d^7	22111	4.00	4.00	4.00	4.00
d^6	22200	5.33	**8.00**	**8.00**	5.75
d^5	11111	4.00	4.00	4.00	4.00
d^0	00000	8.00	8.00	8.00	8.00

[a] This notation represents the occupations of the d orbitals and the most stable orbital is represented on the left

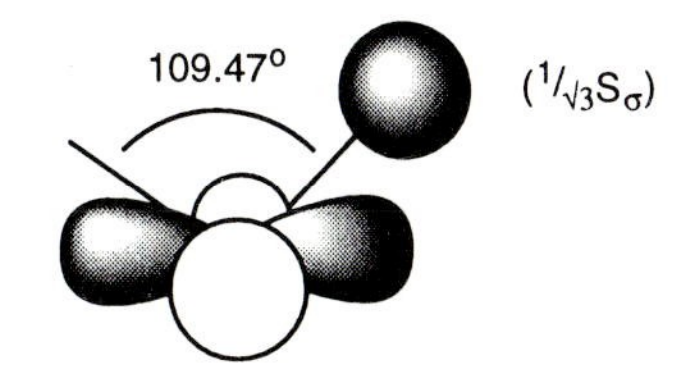

Fig. A.20 Overlap between a ligand orbital and the metal d_{xy} orbital in a tetrahedral complex

In these situations the geometries of the complexes are determined by ligand–ligand repulsion energies. These are minimized for the tetrahedron,

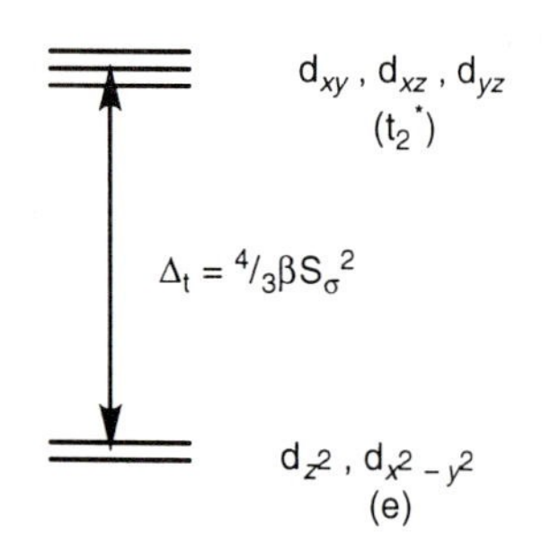

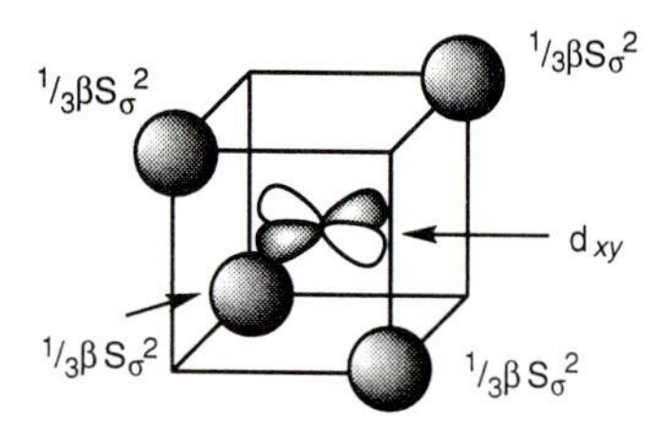

Fig. A.21 Orbital interaction energies and energy diagram in a tetrahedral complex

which places the ligands farthest apart on average. Table A.18 gives examples of specific tetrahedral complexes which are associated with d^0, d^5 (high spin), and d^{10} electron configurations.

Table A.18 Examples of tetrahedral transition metal complexes

d^0	d^5(high spin)	d^{10}
$TiCl_4$	$[MnCl_4]^{2-}$	$Ni(CO)_4$
OsO_4	$[FeCl_4]^-$	$[Cu(NCCH_3)_4]^+$
$[OsO_3N]^-$		$Pt(PMe_3)_4$

It is also apparent from Table A.17 that complexes with (22111) electronic configurations, i.e. with the d_{xz}, d_{yz}, and d_{xy} orbitals in the tetrahedron half filled show no preferences. These complexes are anticipated to be tetrahedral. Cobalt(II) provides many examples of such tetrahedral complexes, e.g. $[CoCl_4]^{2-}$ and $[Co(NCS)_4]^{2-}$.

Low spin d^8 (22220) complexes show a marked electronic preference for square planar geometries and therefore it is not surprising that there are many examples of such complexes and particularly for the second and third row transition metals where βS_σ^2 is larger as a result of greater metal–ligand overlap integrals. For first row transition metals βS_σ^2 is not so large and a fine balance results for tetrahedral and square planar geometries. Since the electronic effects favour a square planar geometry and ligand–ligand repulsions a tetrahedral geometry then increasing the ligand repulsions increases the preference for a tetrahedral geometry. For example $[NiCl_2(PR_3)_2]$ complexes with phosphines with small cone angles favour square planar geometries whereas larger phosphines favour a tetrahedral geometry. Interestingly the d^8 complexes which are square planar universally adopt a low spin configuration because this leads to a larger stabilization energy. Therefore the geometric transformation from square planar to tetrahedral has associated a change of spin state since the latter has two unpaired electrons in the t_2 (d_{xz}, d_{yz}, and d_{xy}) orbitals.

Given that tetrahedral complexes are favoured for d^{10} and square planar for d^8 then it will not come as a great surprise that d^9 complexes show a wide range of geometries which lie between these two geometric ideals.

For low spin d^6 complexes (22200) the square planar and folded square geometries have equal energies and both are significantly more stable than either the tetrahedron or the trigonal pyramid. The experimental structural data which are available suggest that the folded square is the more stable geometry. The analysis provided above has ignored π-bonding effects—if these are taken into account then the folded square which is able to utilize the d_{xz}, d_{yz}, and d_{xy} orbitals more effectively in π-bonding leads to a more stable geometry. $Cr(CO)_4$, which has been studied in inert gas matrices, provides a specific example of such a d^6 complex.

Table A.19 summarizes the d orbital splittings for other common coordination geometries calculated using the same principles as those developed above.

Table A.19 Energies of the five d orbitals in complexes of various geometries and coordination numbers and their total energies in units of βS_σ^2

Geometry	d_{z^2}	$d_{x^2-y^2}$	d_{xz}	d_{yz}	d_{xy}	Total energy in metal d orbitals
Tricapped trigonal prism	−1.2796	−1.6191	−2.2411	−2.2411	−1.6191	−9.0000
Square antiprism	0.0000	−1.3333	−2.6667	−2.6667	−1.3333	−8.0000
Dodecahedron	−1.3873	0.0000	−1.8551	−1.8551	−2.9026	−8.0000
Bicapped trigonal prism	−1.0296	−1.4316	−2.2411	−2.2411	−1.0566	−8.0000
Pentagonal bipyramid	−3.2500	−1.8750	0.0000	0.0000	−1.875	−7.0000
Capped trigonal prism	−0.7796	−0.6816	−2.2411	−2.2411	−1.0566	−7.0000
Trigonal prism	−0.5296	−0.4941	−2.2411	−2.2411	−0.4941	−6.0000

These data are useful for extending the geometric analyses to other coordination numbers. They may also be used for developing broader generalizations. For example, the d orbital splittings shown in Fig. A.22 provide an insight into the electronic factors responsible for the Effective Atomic Number (EAN) Rule. The sequential increase in the number of non-bonding orbitals localized on the metal as the number of ligands decreases from 9 to 6. For lower coordination numbers the adherence to the EAN rule may only be achieved if orbitals which are antibonding along the metal-ligand bonds are occupied. These antibonding effects are mitigated by π-back donation effects and d–p orbital mixings. The higher lying p orbitals are not considered in the angular overlap simplification but for the trigonal bipyramid, the tetrahedron and the trigonal plane the some or all of the p orbitals match the symmetries of the metal d orbitals which are responsible for the antibonding interactions, and d–p orbital mixings reduce their antibonding character.

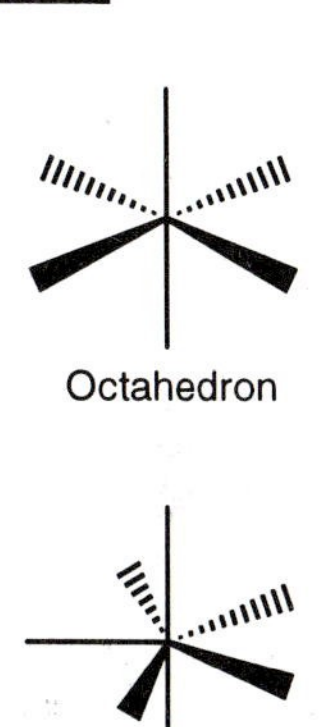

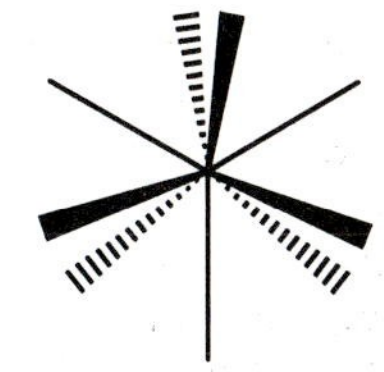

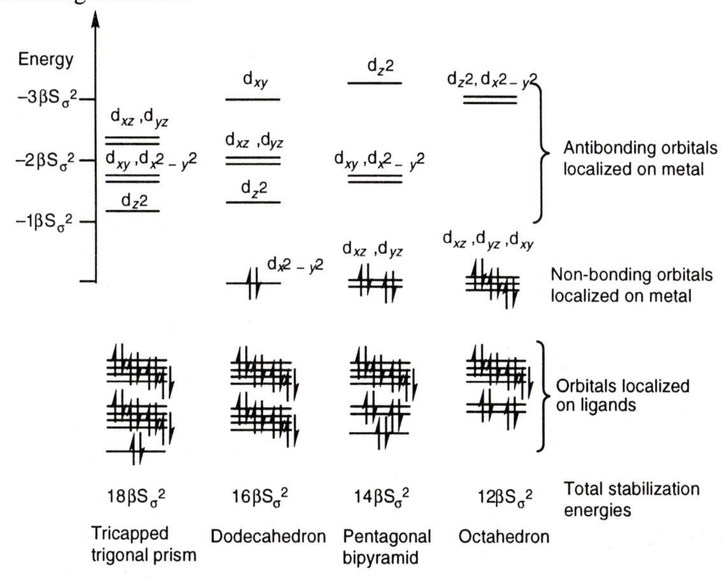

Fig. A.22 Evolution of molecular orbitals for 9 – 6 coordinate complexes which illustrates the application of the effective atomic number (18 electron) rule

B

Band theory

Band theory

Band theory extends the molecular orbital theory ideas which were developed for simple molecules to infinite solids. It is particularly useful for discussing the electrical conductivity and magnetic properties of compounds where the interactions between the individual components, be they atoms or molecules, are significant. It is therefore useful to briefly review the basic concepts underlying a molecular orbital analysis of the bonding in a simple diatomic molecule derived from second row atoms, e.g. B_2, C_2, N_2, O_2. As the two atoms of the diatomic molecule which are initially a long distance apart are brought together molecular orbitals which are delocalized over the whole molecule are generated by the overlap of symmetry-matching atomic orbitals. The 2s–2s, $2p_\sigma$–$2p_\sigma$, and $2p_\pi$–$2p_\pi$ overlaps which fall into this category are shown in Fig. B.1. The resulting molecular orbitals at the equilibrium distance are illustrated in Fig. B.2. Since there are initially 8 valence atomic orbitals 8 molecular orbitals are generated and their symmetries are illustrated in Fig. B.2. The relative energies of the orbitals are determined primarily by the following parameters:

1. The efficiency of orbital overlap between the two atoms, i.e. 2s–2s, $2p_\sigma$–$2p_\sigma$, and $2p_\pi$–$2p_\pi$. The larger the overlap the greater is the energy difference between the bonding and antibonding combinations, σ_g–σ_u*, π_u–π_g*.

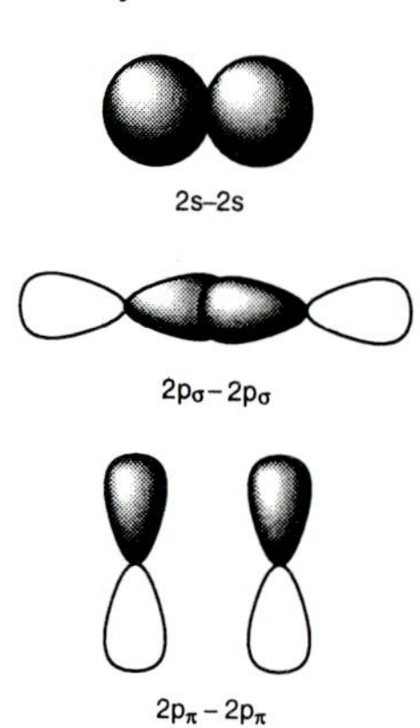

Fig. B.1 Pairwise overlap between the valence orbitals in a diatomic molecule

The electronic configurations of the second row diatomic molecules are given below. They all possess the core $1\sigma_g^2 1\sigma_u^2$ configuration.

B_2	$1\pi_u^2$		
C_2	$1\pi_u^4$		
N_2	$1\pi_u^4$	$2\sigma_g^2$	
O_2	$2\sigma_g^2$	$1\pi_u^4$	$1\pi_g^2$
F_2	$2\sigma_g^2$	$1\pi_u^4$	$1\pi_g^4$

The s–p mixing effects account for the orbital occupations in the ground states of the diatomic molecules B_2 – N_2. For O_2 and F_2 the s–p mixing is reduced because of the greater s–p separation and the orbital occupation of $2\sigma_g$ and $1\pi_u$ are reversed.

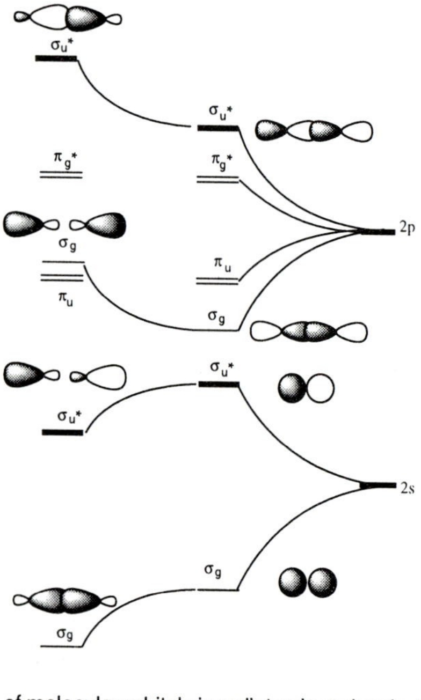

Fig. B.2 The formation of molecular orbitals in a diatomic molecule. On the right-hand side the energies of valence orbitals of the isolated atoms are shown. The energy levels in the centre indicate the molecular orbitals which result if the interatomic orbitals are taken in a pairwise fashion s with s and p with p. The effect of s–p mixing is shown on the left-hand side. Molecular orbitals with matching symmetries repel each other

2. The energies of the s and p orbitals in the isolated atom are important because the stabilization energies discussed in 1 are referenced with respect to the s and p orbital energies.

The resulting molecular orbital diagram of Fig. B.2 has ignored the possibility of overlap between an s orbital on one atom and a p orbital on the second and vice versa, which is also symmetry allowed. The extent of interaction depends on the overlaps involved and the 2s–2p orbital separation. This supplementary interaction is introduced into the molecular orbital scheme by allowing molecular orbitals of the same symmetry to interact in such a way that the bonding component is stabilized and the antibonding component is destabilized. These interactions are illustrated on the left-hand side of Fig. B.2 and the atomic orbital overlaps contributing to the mixing of s and p character are shown in Fig. B.3.

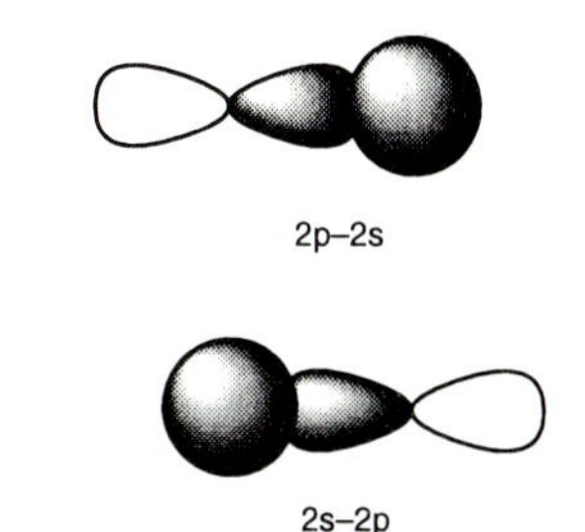

Fig. B.3 s–p orbital overlap responsible for s–p mixing in diatomics

This relatively simple molecular orbital analysis is useful because it provides a rationalization of the following experimental features.

1. The ground state configurations of the molecules which result from the *aufbau* filling of the molecular orbitals. The correct assignment of the ground state electronic configurations requires that the s–p orbital mixings discussed above are taken into account.
2. The dissociation energies of the diatomic molecules shown in Fig. B.4 reach a maximum at N_2. For this molecule the molecular orbitals generated from the p shell are precisely half filled with the σ and π bonding components contributing to a maximum bond order of three.

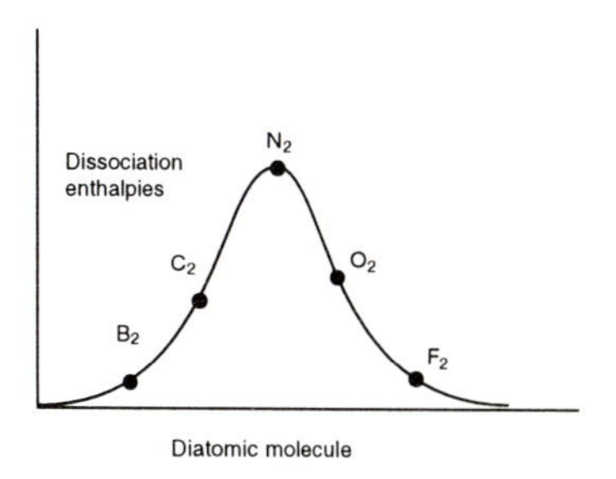

Fig. B.4 Bond dissociation enthalpies in diatomic molecules of the second row atoms

These ideas may be extended to an infinite solid where all the atoms are arranged in a closed packed manner with the following modifications. If a large number of atoms are brought together a very closely spaced set of molecular orbitals is generated rather than a few well spaced orbitals. Fig. B.5 illustrates the evolution of molecular orbitals as the number of atoms is increased. This simple picture is based on a string of alkali metal atoms. The resultant spectrum of closely spaced molecular orbitals is described as a *band*. The width of the band of molecular orbitals is decided primarily by the extent of overlap between the orbitals. Very efficient overlap between the orbitals leads to a large band width, whereas ineffective overlap leads to a narrow band.

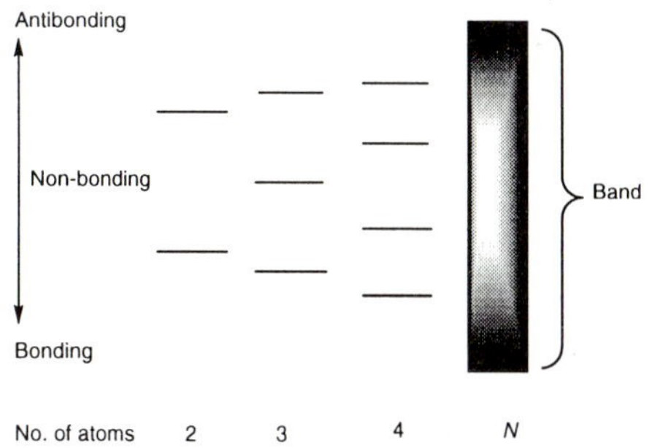

Fig. B.5 The evolution of a band of molecular orbitals as the number of contributing atoms increases

Fig. B.5 suggests that each energy state is associated with only a single molecular orbital, but this is an oversimplification. If the string of alkali metal atoms is studied in more detail it is apparent that there is only one combination of s orbitals which is totally in phase. This is the most stable molecular orbital. Similarly there is only one combination that is totally out-of-phase and this is the least stable molecular orbital. However, for molecular orbitals lying between these extremes, there are a number of permutations of in- and out-of-phase orbitals which have identical energies. For example, there are the two equivalent non-bonding combinations shown in the middle of Fig. B.6. They have equal energies but the precise location of the nodes differs. Molecular orbitals which lie between the non-bonding levels and the bottom and top of the band have many more equivalent permutations. Therefore, the bands are better represented by the line diagram shown on the left in Fig. B.7.

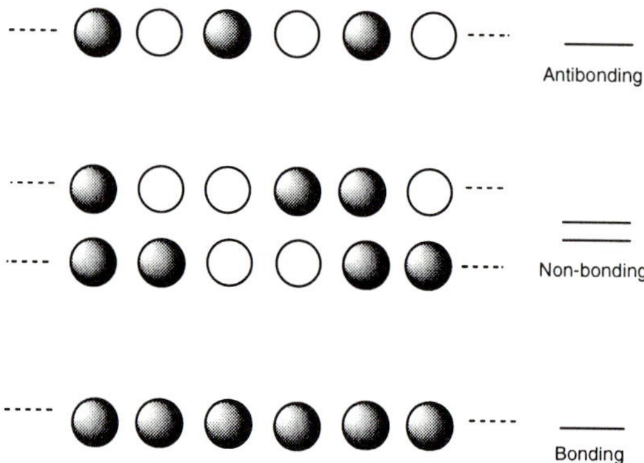

Fig. B.6 The phase relationships in a band of molecular orbitals. Note the most stable orbital has no nodes and the least stable a node between each atom. There are two linear combinations which give rise to non-bonding molecular orbitals

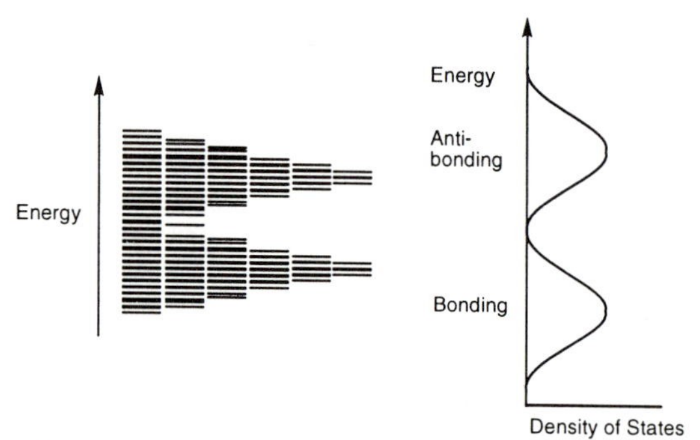

Fig. B.7 For an infinite array of atoms the band structure is better represented as a mathematical function than a histogram

Since the number of atoms in a typical metal approximates to an infinite number the histogram representing the energy levels may be replaced by a density of states diagram such as that shown on the right-hand side of Fig. B.7. The horizontal axis represents the number of energy levels between E and $E + \delta E$. The number of states within each band is equal to the total of all orbitals contributed by the N atoms. For example, an s band will consist of N states and a p band $3N$ states. Each energy state may be occupied by a pair of electrons with opposing spins, conforming to the Pauli Exclusion Principle, and the states are filled in an *aufbau* fashion. The bands will generally contain the electrons that resided in the parent atomic orbital, and this can result in empty, partially filled, or totally filled bands. In addition the bands can overlap in much the same way that the s and p atomic orbitals mix in diatomic molecules.

The extent to which the bands remain distinct depends on the s–p orbital separation and the overlaps—if the s–p separation is large and the overlaps small then the bands remain distinct (Fig. B.8(a)). However, if the s–p separation is smaller and the overlaps are large then extensive band mixing occurs (Fig. B.8(b)). For the alkaline earth metals and the metals of the main group elements the extent of overlap between the s and p bands is large, i.e. the situation is analogous to that shown on the right-hand side of Fig. B.8(b).

The electronic properties of a solid are closely connected with the band structure. The band that contains the highest energy electrons is called the valence band and the first unoccupied energy levels above these are described as the conduction band.

There are four basic band structure types and they are shown in Fig. B.9. In (a) the valence band is only partially occupied by electrons. The energy corresponding to the highest occupied state at 0 K is described as the Fermi energy, E_f.

Metals such as sodium and copper which have a half-filled s band are well represented by such a diagram. Alkaline earth metals and zinc, cadmium and mercury have just the right number of electrons to fill an s band. However, s–p band mixing of the type illustrated in Fig. B.8 occurs and the resultant combined band structure remains incompletely filled and these elements are metallic conductors. The diagram in Fig. B.9(b) is a reasonable representation of this situation. Band theory always associates conductivity with the presence of partially filled bands.

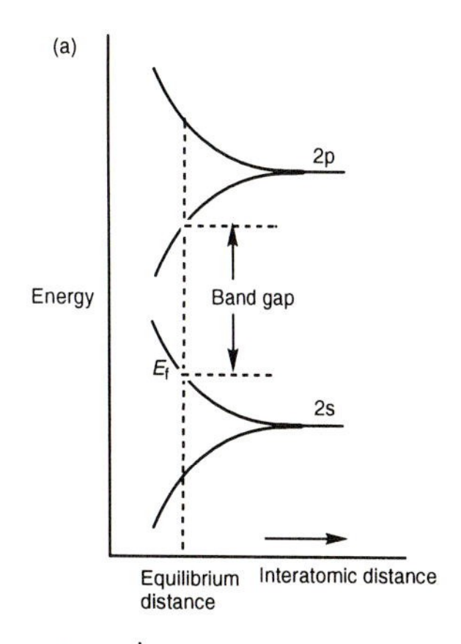

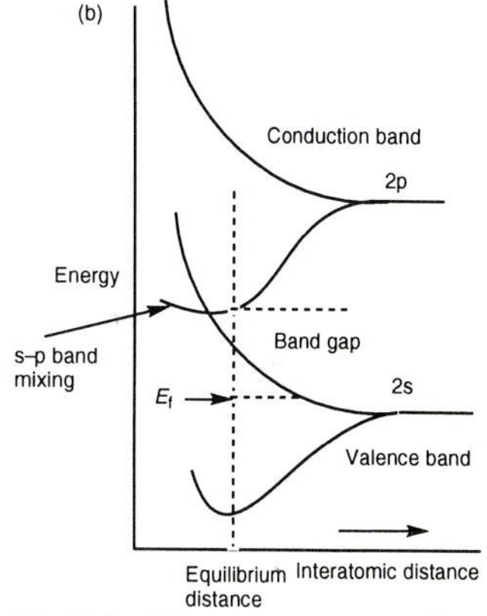

Fig. B.8 Schematic representations of the development of a band structure (a) in the absence of extensive s–p mixing and (b) with s–p mixing

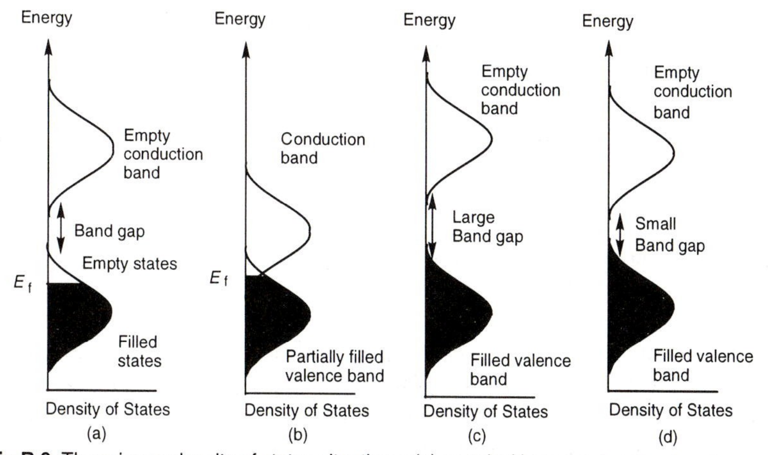

Fig B.9 The primary density of states situations; (a) metal with no overlapping bands, (b) metal with overlapping bands, (c) an insulator, and (d) a semiconductor

If the valence states are completely filled and the conduction band is empty then the solid has poor electronic conductivity properties. If the band gap is large (Fig. B.9(c)) the solid is an insulator, whereas if the gap is smaller the solid is a semiconductor (Fig. B.9(d)).

The relationship between the band structure and electrical conductivity needs some further amplification. In order for an electron to become sufficiently free to be accelerated by an external electric field it has to have empty energy states very close to the level associated with the Fermi energy. Once an electron is promoted to these levels it experiences much less attraction from adjacent nuclei and can rapidly migrate through the solid when the electric field is applied. Therefore, conduction can be associated either with a partially filled valence band or the presence of a very small band gap between the valence band and the conduction band. All metals and alloys have partially filled valence bands. Graphite provides an interesting example of a conducting material which has a filled valence band but a zero band gap between the Fermi level and the conduction band.

The resistivity of a metal is influenced by imperfections in the crystal lattice due to impurity atoms, vacancies, interstitial atoms, dislocations, and the thermal vibrations of the atoms. All of these scatter the electron and make it change its direction. The mobility of the electron through the lattice is therefore an important parameter in influencing its resistivity. The number of electron carriers, i.e. the number of free electrons, is also important. The resistivity of a metal increases with temperature primarily because of the increase in the number of defects and the greater amplitudes of motion of the metal atoms in the lattice.

Figs. B.9(a) and B.9(b) illustrate band structures which are appropriate to metals and are associated with partially filled band structures. The band structures in Figs. B.9(c) and B.9(d) represent situations where the valence band is filled and there is a significant band gap to the conduction band. If the band gap is greater than 3 eV then the material is an insulator. Diamond would provide a good example of such a situation. However, if the band gap is less than 3 eV the pure material is described as an intrinsic semiconductor. The conductivity of a semiconducting material increases with temperature because more electrons will have sufficient energy to be promoted into the conduction band. The conductivity also increases as the band gap decreases, For example, Ge which has a band gap approximately half that of Si has a conductivity four orders of magnitude larger.

The band gap depends not only on the chemical composition of a material but also its physical state. For example, an amorphous material has a smaller band gap than a crystalline material.

It should be emphasized that effective conduction is achieved in a semiconductor even when relatively few electrons are promoted. For example in pure silicon it has been estimated that there are approximately 1×10^{16} charge carriers per cubic metre where the total number of atoms is 5×10^{28}. It is important to note that the promotion of an electron from the valence band into the conduction band not only creates a charge carrier in the conduction band but also leaves a hole in the valence band. This hole provides a vacancy which can promote the movement of electrons in the conduction band. The movement of an electron into this hole creates another hole where the electron originally resided. The situation resembles a solitaire board. If all the holes are filled by pegs then it is impossible to start the game, but once a single hole is created at the centre there are a number of moves which allow the pegs to be moved around the board. As the game proceeds the number of possibilities for moving the pegs increases rapidly and it is this which provides the challenge to the player wishing to plot the correct sequence of moves to leave only a single peg. The conductivities of semiconductors are also very sensitive to the presence of impurities. Where an impurity leads to either an occupied level in the band gap just below the conduction band or a vacant level just above the valence band such materials are described as extrinsic semiconductors.

Table B.1 Conductivities of some metals, semiconductors, and insulators:

Example		conductivity $(\Omega m)^{-1}$	Band gap (eV)
Metals:			
	Copper	6.0×10^7	0
	Sodium	2.4×10^7	0
	Magnesium	2.2×10^7	0
	Aluminium	3.8×10^7	0
Zero band gap semiconductor:			
	Graphite	2×10^5	0
Semiconductors:			
	Silicon	4×10^{-4}	1.11
	Germanium	2.2	0.67
	GaAs	1.0×10^{-6}	1.42
Insulators:			
	Diamond	1×10^{-14}	5.47
	Polythene	10^{-15}	

Table B.1 gives a summary of the electrical conductivities of some solids which fall into the four categories illustrated in Fig. B.9.

The band model developed above is useful for rationalizing the relative enthalpies of atomization of the elements across rows of the periodic table (See Fig. B.10). It is noteworthy that for each row the maximum occurs for the Group 14 element which has four valence electrons. This is just the appropriate number for half filling the band structure which develops from the presence of four s and p orbitals per atom. Interestingly the trend appears to be insensitive to whether the element forms a metallic, infinite covalent, or molecular structure in the solid state. The other significant feature of Fig. B.10 is that in general the heats of atomization fall on descending a column of the periodic table and this reflects the weaker element–element bonding down a column as the atomic number increases.

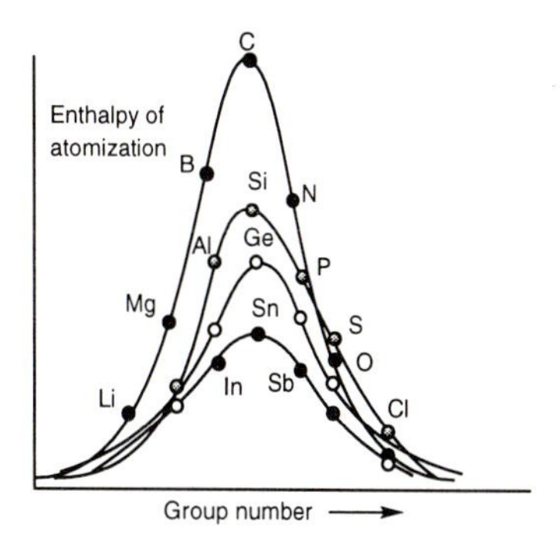

Fig. B.10 Enthalpies of atomization for the main group elements

The data shown in Table B.2 for the fourth row elements, which include the first transition series, suggest that in general the melting points, boiling points, and enthalpies of atomization of these elements are closely correlated. The s and d block elements (K to Zn) show an overall increase in $\Delta_{at}H^{\ominus}$ up to V and then there is a marked decrease to zinc. The $\Delta_{at}H^{\ominus}$ values for Cr and

Mn are anomalously low for reasons associated with the respective $4s^1 3d^5$ and $4s^2 3d^5$ configurations of these atoms.

Table B.2 Melting and boiling points and standard enthalpies of atomization for the third row elements

	K	Ca	Sc	Ti	V	Cr	Mn	Fe	Co
m.p./ °C	64	845	1539	1675	1900	1890	1244	1535	1495
b.p./ °C	774	1487	2727	3260	3400	2480	2097	3000	2900
$\Delta_{at}H^{\ominus}$ / kJ mol^{-1}	90	177	390	469	502	397	284	406	439
	Ni	Cu	Zn	Ga	Ge	As	Se	Br	
m.p./ °C	1453	1083	419	30	937	817	217	–7	
b.p./ °C	2732	2595	907	2403	2830	subl.	685	59	
$\Delta_{at}H^{\ominus}$ / kJ mol^{-1}	427	341	130	277	376	287	207	111	

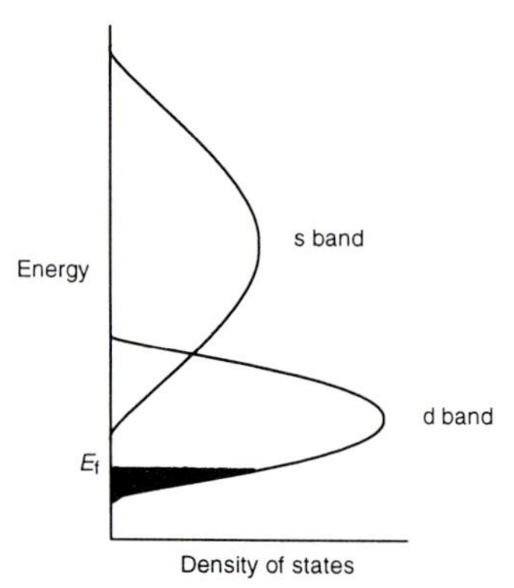

Fig. B.11 Density of states diagram for a transition metal

Aufbau filling of d band with the spins pairing up.

Non-*aufbau* filling of a d band. Some of the spins remain parallel leading to magnetic properties and occupation of higher energy and more anti-bonding levels in the band. This results in a weakening of the metal–metal bonding.

Fig. B.12 *Aufbau* and non-*aufbau* fillings of metal d band orbitals

For the transition metals the band structure is dominated by overlapping d and s bands as shown in Fig. B.11. If the d band of molecular orbitals is filled in an *aufbau* fashion the $\Delta_{at}H^{\ominus}$ should increase from a low initial value for d^0 to a maximum at d^5 (the half-filled shell), and a minimum for the completely filled shell (d^{10}) . The s orbitals also generate a band which gives maximum binding energies for s^1 configurations. Therefore, for the transition elements the maximum binding energies (and $\Delta_{at}H^{\ominus}$ values) should be achieved for the $4s^1 3d^5$ configuration and minimum for the $4s^2 3d^{10}$ configuration (if d–s orbital mixing effects are neglected).

Although zinc does have the smallest $\Delta_{at}H^{\ominus}$, the maximum value is actually achieved for vanadium which has five valence electrons and then dips at chromium and manganese. For the transition elements the 3d electrons have particularly high electron–electron repulsion energies and consequently the idealized *aufbau* filling of the band is not always achieved. Some of the molecular orbitals which are metal–metal bonding are only occupied by single electrons and this forces the additional electrons to occupy more antibonding metal–metal orbitals. Although the occupation of these orbitals reduces the metal–metal bonding, their occupation is favoured because the electron–electron repulsion effects are less than if the electrons were forced to occupy a bonding orbital which already contains a single electron.

Such elements therefore have lower than anticipated $\Delta_{at}H^{\ominus}$ values. These *aufbau* and non-*aufbau* fillings of metal orbital bands are illustrated schematically in Fig. B.12. The large exchange energies associated with the $4s^1 3d^5$ (Cr) and $4s^2 3d^5$ (Mn) configurations lead these elements to have non-*aufbau* configurations when the metallic band structures are populated and therefore they have the anomalously low $\Delta_{at}H^{\ominus}$ values shown in Fig. B.13.

Fig. B.14 illustrates the $\Delta_{at}H^{\ominus}$ values for the second and third row transition elements respectively. For these elements the larger sizes of the 4d and 5d orbitals (as compared with those of the 3d orbitals) reduces the extent of electron–electron repulsion and the filling of the band structures occurs in a more *aufbau* fashion. The $\Delta_{at}H^{\ominus}$ values consequently follow the theoretical curve more closely. For the second row the maximum occurs at Nb (5 valence electrons) and W (6 valence electrons) for the third row and only a small dip in the graph is observed for the next metal. It is also noteworthy that the enthalpies of atomization follow the order: 3rd row > 2nd row > 1st row. These differences reflect the relative overlaps between the metal d valence orbitals, i.e. 5d–5d > 4d–4d > 3d–3d.

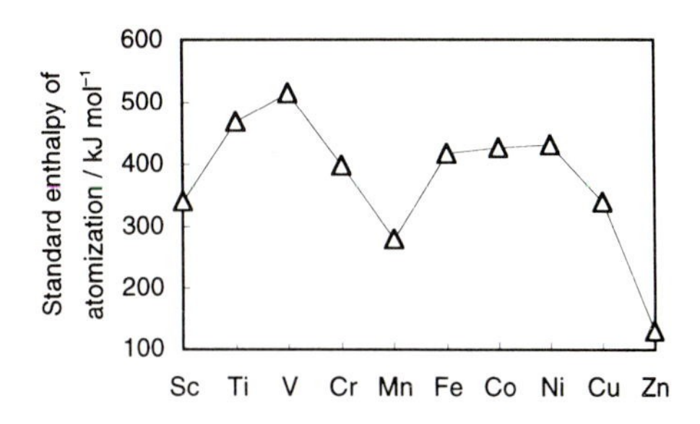

Fig. B.13 Standard enthalpies of atomization of the first row transition elements, Sc–Zn

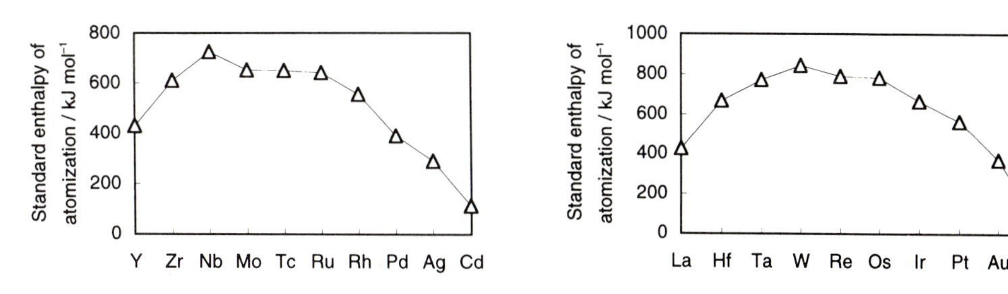

Fig. B.14 Standard enthalpies of atomization of the second (Y–Cd) and third (La–Hg) row transition elements

Metallic radii and densities

Across a row of the Periodic Table the sizes of the atoms generally decrease because the effective nuclear charge of the atoms increase. The data given in Table B.3 confirm this trend is reproduced for the metallic radii of the pre-transition elements and the transition elements, with the exception of manganese and copper which have larger radii than anticipated.

As noted above the electron–electron repulsion energies for the d electrons of the first row transition elements are large and therefore filling of the band structure representing the delocalized molecular orbitals in a metal does not always follow a simple *aufbau* procedure. The respective $4s^2 3d^5$ and $4s^1 3d^{10}$ electronic configurations of manganese and copper and their associated *exchange energies* result in the population of more antibonding sections of the band structure and longer metal–metal bonds result.

The decreasing radii and increasing relative atomic masses lead to an increase in densities across the series.

Table B.3 Metallic radii and densities of the elements K–Se of the fourth row of the Periodic Table

	K	Ca	Sc	Ti	V	Cr	Mn	Fe
Metallic radius/ pm	235	197	164	147	135	130	135	126
Density / kg m^{-3}	860	1540	3000	4500	6100	7200	7440	7860
	Co	Ni	Cu	Zn	Ga	Ge	As	Se
Metallic radius / pm	125	125	128	137	141	137	139	140
Density / kg m^{-3}	8860	8900	8920	7130	5910	5320	5730	4790

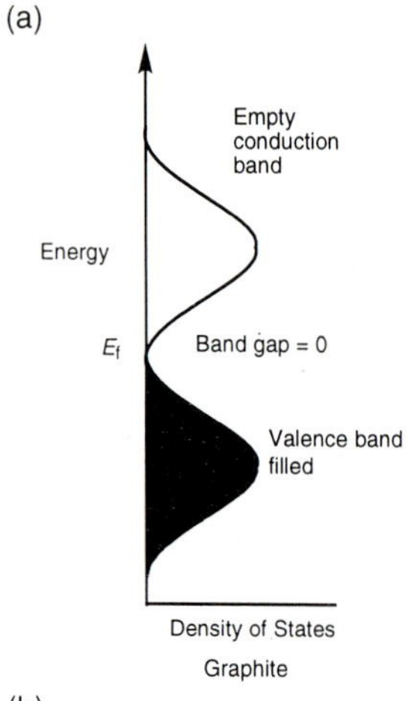

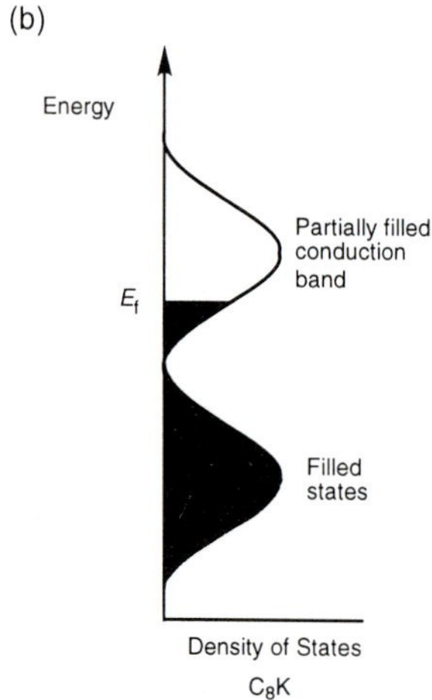

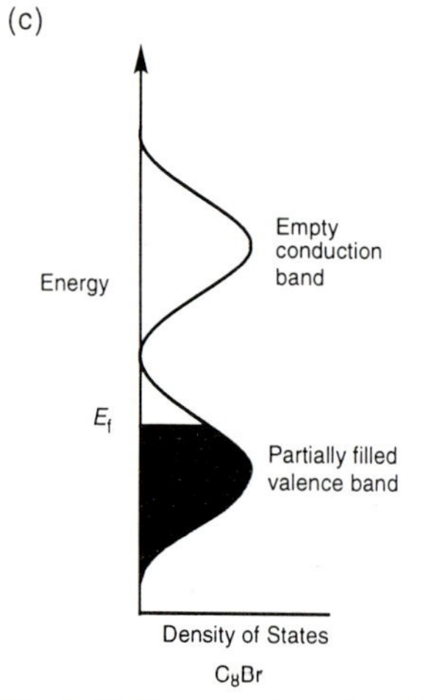

Fig. B.15 Band structures for graphite and its intercalates

Other examples of band theory applications

Graphite

The band structure of graphite indicates that it is a zero band gap semiconductor, i.e. the top of the band resulting from the bonding π–molecular orbitals just touches the bottom of the band arising from the antibonding π^* molecular orbitals (see Fig. B.15(a)). This gap is significantly smaller than thermal energies and therefore a few electrons are excited from the valence band into the conduction band and graphite therefore has a significant conductivity.

If electrons are removed or added to the band structure associated with graphite the Fermi level moves down into the valence band or up into the conduction band and there are now many levels accessible to the electrons thermally.

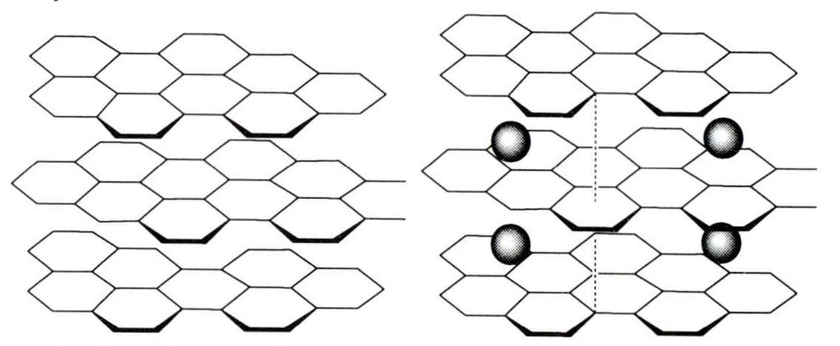

Fig. B. 16 Schematic illustration of the intercalation process for graphite

In chemical terms this is achieved by intercalating ions between the layers of carbon atoms in graphite. The intercalation process is illustrated schematically in Fig. B.16. For example, the reaction of graphite with potassium gives C_8K, which in view of the very electropositive nature of potassium is better represented as $C_8^-K^+$. This means that one eighth of an electron per carbon enters into the conduction band and the density of states resembles that shown in Fig. B.15(b). Addition of an oxidizing agent such as bromine results the intercalation of Br^- ions between the graphitic layers and the formation of a compound with the formula $C_8^+Br^-$ and the loss of one eighth of an electron from the valence band of graphite as shown in Fig. 15(c). In contrast to graphite itself both of these intercalated derivatives are genuine metals and their conductivities are orders of magnitude greater than that of the parent graphitic material.

Transition metal oxides

The metal oxides, MO, generally have the sodium chloride structure although at times they are not simple stoichiometric compounds. In the sodium chloride structure the octahedral environment around the metal leads to the usual e_g–t_{2g} splitting (see Angular overlap model). The non-bonding t_{2g} set point towards the matching orbitals on the adjacent metal ions and the overlap between the t_{2g} orbitals is small but sufficiently significant for a band structure to develop. The band structure is wider for the earlier metals in the transition series because the effective nuclear charge they experience is smaller and the orbitals are larger and overlap more effectively. These effects are illustrated for TiO and NiO in Figs B.17 and B.18. The resultant band is able to accommodate a total of 6 electrons and consequently for an ion such as Ti^{2+} it is partially filled and the resultant oxide is an electrical conductor whose conductivity equals $10^3\ (\Omega cm)^{-1}$.

The other early transition metal oxides are also conductors because the band width is sufficiently large to give rise to effective delocalization of the electron density through the lattice and the bands are partially filled. For Ni^{2+} however, the band width is much narrower (smaller orbital overlaps) and the t_{2g} band is completely filled. The e_g orbitals which contain two electrons for the d^8 Ni^{2+} ion also give rise to a band structure, but since these orbitals point at the oxide anions rather than adjacent metals they do not create a band which is effectively delocalized over all the metal atoms. Therefore, NiO shows no metallic conduction properties and behaves as an insulator.

The metal trioxides of the transition metals also display some interesting conducting properties. WO_3 is an insulator and is pale yellow in colour. It has the ReO_3 structure based on vertex sharing octahedra shown in Fig. B.19. structure creates a cavity at the centre of the eight octahedra arranged at the vertices of a cube. In these materials the metals have octahedral coordination geometries and the d orbitals split into the usual e_g and t_{2g} sets. However, in the ReO_3 structure the metals are too far apart to overlap directly. The t_{2g} orbitals do overlap strongly with the p orbitals of the oxide ions to generate a band structure with a significant width. The relevant orbital interactions are illustrated in Fig. B.20.

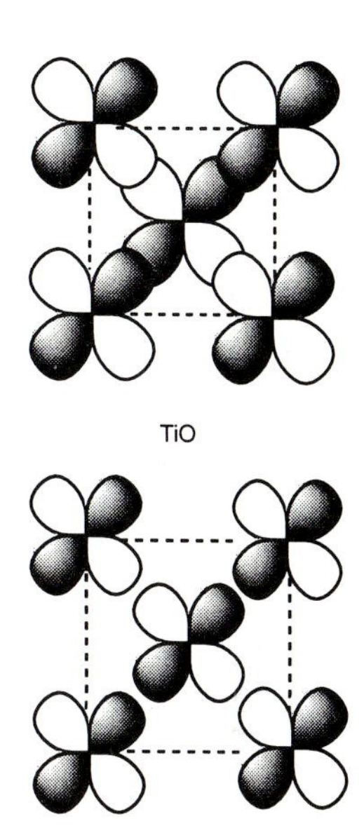

Fig. B.17 A diagram illustrating the differences in overlap between metal d orbitals in TiO and NiO

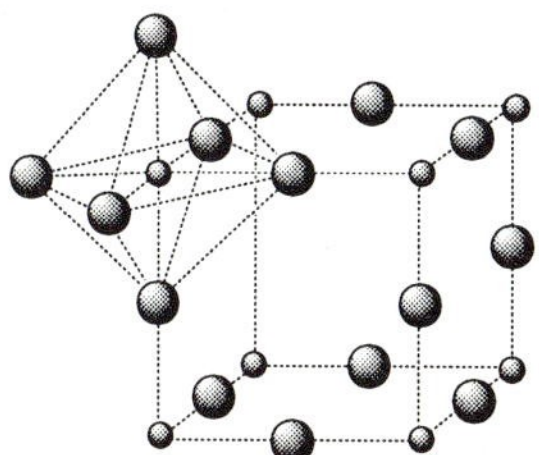

Fig. B.19 ReO_3 structure showing vertex sharing of octahedra and the central cubeoctahedral cavity

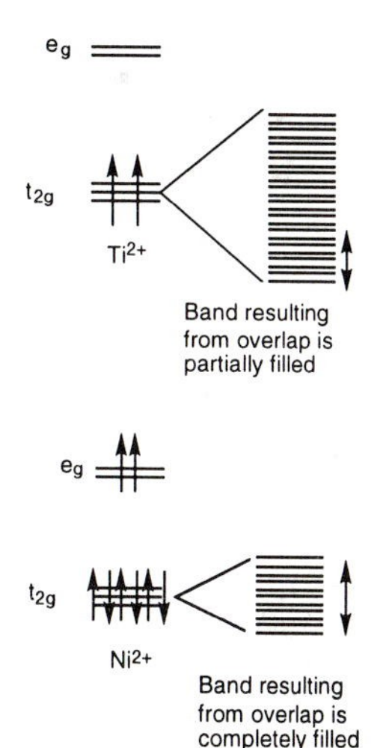

Fig. B.18 d-band formation and occupation in Ti^{2+} and Ni^{2+}

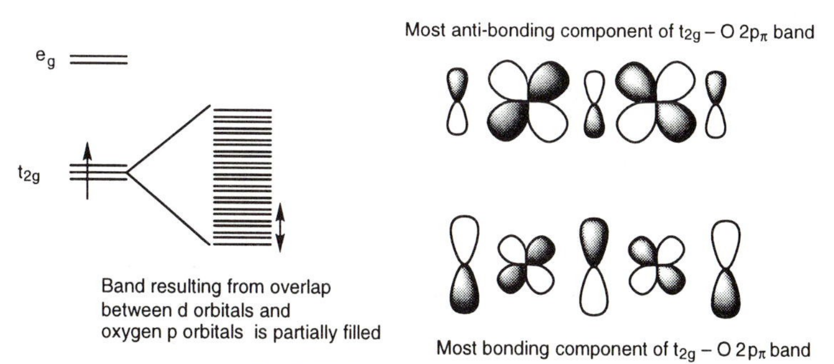

Fig. B.20 Band structure for ReO_3

The bonding component of the band is localized primarily on the oxygen atoms and the antibonding component is localized mainly on the metal atoms. In WO_3 ($W^{VI} - d^0$) only the bonding component is filled and there is a significant band gap between this filled valence band and the empty conduction band and therefore it is an insulator. In contrast ReO_3 (Re^{VI} - d^1) has a single electron occupying the band and therefore it is a metallic conductor. The reduced compound $Na_{0.6}WO_3$ is also conducting because the electrons contributed by the sodium atoms enter into this band. The sodium ions are accommodated within the cuboctahedral cavity of the ReO_3 structure. These compounds are described as tungsten bronzes because of their appearance.

Photoconductivity

The discussion of semiconductors given above has concentrated around the excitation of electrons into the conduction band by thermal effects. They can also be excited by the absorption of light giving rise to the phenomenon of photoconductivity. For example, the visible spectrum of GaAs shows a distinct maximum at 1.5 eV (cf. the band gap of 1.42 eV given in Table B.1) which can be associated with such an excitation. The promotion of electrons into the conduction band creates pairs of electrons and holes which are held together by their mutual attraction. The application of an electric field causes them to migrate in opposite directions and conduction results. Photoconducting materials are widely used in the photocopying process.

Grey selenium has a structure based on helical chains of atoms and in a crystalline form it is a semiconductor with a band gap of 2.6 eV. The band gap is reduced in amorphous forms to about 1.8 eV. Selenium is laid down to a thickness of 50 mμ on an aluminium metal plate by a vapour deposition process and an electrostatic positive charge is induced by a corona discharge on to the plate. In the dark the semiconductor has a poor conductivity and the charge is not dissipated. When light falls on the plate in a fashion which reflects the white and black parts of the paper being photocopied then those areas of selenium exposed to the more intense light become photoconductive. The charge is discharged only perpendicular to the surface and therefore the image is recreated in terms of areas of charged and uncharged selenium particles. The powdered ink adheres to those parts of the plate which have retained an electrostatic charge. The powdered ink is transferred to a fresh piece of paper and the image fixed by heating.

Doped semiconductors

It is noted earlier that pure silicon is a semiconductor. When phosphorus is doped into the silicon lattice the phosphorus atom can occupy some of the tetrahedral sites initially occupied by silicon. The valence band of silicon is full and therefore the additional valence electron of phosphorus must go somewhere. To a first approximation one can say the additional electron just goes into the conduction band, but this is an over-simplification. The phosphorus atom which now has a formal charge of +1 exerts a stronger attractive force towards the electron than silicon and therefore this electron is slightly less free than the conduction electrons around silicon and therefore a donor state is created just below the conduction band as illustrated in Fig. B.21(b).

Alternatively an aluminium atom could be introduced into the silicon structure and occupy a tetrahedral site. Since aluminium has one electron fewer than silicon a hole in the valence band is created. Again the hole does not occur in the centre of the valence band, but slightly above it because formally the aluminium has been substituted as Al^-. This acceptor level is illustrated in Fig. B.21(c). For narrow band semiconductors such as Si, Ge, GaAs, and PbTe the donor and acceptor levels created in this manner are very close to the edges of the bands and therefore can participate very effectively in conduction processes. In many other semiconductors which have lower dielectric constants the situation is less favourable for conduction.

The electronic properties of semiconductors are determined by the concentrations of the charge carriers, i.e. the electrons and the holes. In an intrinsic semiconductor the number of holes and electrons is exactly equal, but in extrinsic semiconductors the introduction of impurities leads to an excess of either donor or acceptor sites. *n*-type semiconductors have extra electrons provided by donor levels and *p*-type semiconductors have extra holes originating from the acceptor levels.

For a more comprehensive discussion of band theory see P. A. Cox, *The Electronic Structure and Chemistry of Solids*, OUP, Oxford, 1987

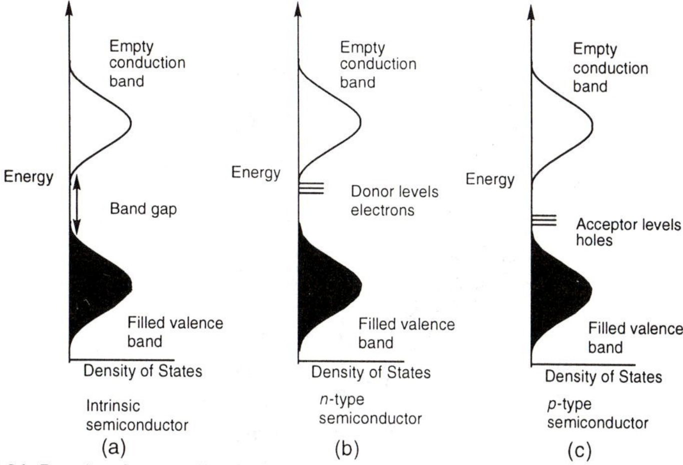

Fig. B.21 Density of states diagrams for (a) an intrinsic semiconductor, (b) an *n*–type semiconductor, and (c) a *p*-type semiconductor

The *p–n* junction is fundamental to the working of solid state electronic devices. These junctions are made by diffusing a dopant of one type into a layer of a semiconductor of the other type. Electrons migrate from the *n* type regions to the *p* type regions forming a space charge region where there are no carriers. The unbalanced charge of the ionized impurities causes the bands to bend as shown in Fig. B.22 until a point is reached where the Fermi levels are equivalent.

The most important property of a *p–n* junction is that of rectification, i.e. current is much easier to pass in one direction than the other. The depletion of carriers for the junction effectively forms an insulating barrier. If a positive potential is applied to the *n* type side the situation known as reverse bias more carriers are removed and the barrier becomes wider. However, under forward bias the *n* type is made more negative relative to the *p* side and the energy barrier is decreased and so the carriers may flow through.

Electrons passing into *p* type semiconductor recombine with holes and the holes vice versa on the other side.

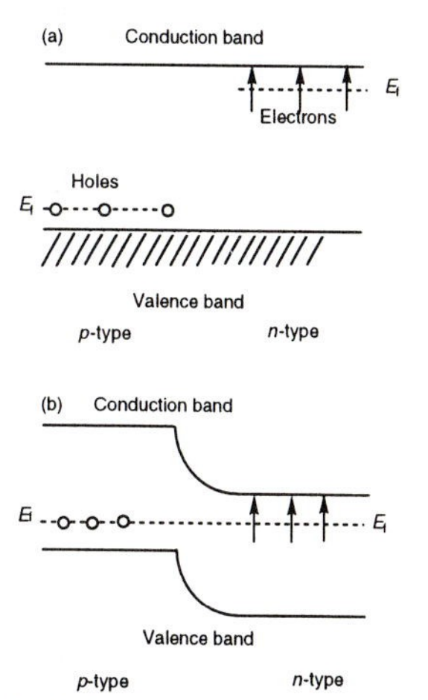

Fig. B.22 A schematic illustration of a *p–n* junction

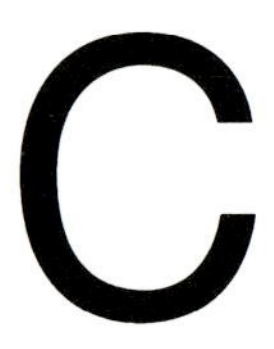

Crystal defects

Examples of compounds showing Schottky defects: MgO, CaO, and alkali metal halides

Examples of compounds showing Frenkel defects: CaF_2, SrF_2, AgCl, AgBr, β-AgI, UO_2, and ZrO_2

Colour centres
F-centres
Colour centres generally result from an electron trapped in an anion vacancy in alkali or alkaline earth metal halides. They may be formed by irradiation of a crystal with X or γ-rays, or neutrons, or by heating the crystal in the vapour of an alkali or alkaline earth metal. The metal atoms diffuse into the crystal and occupy cation sites and an equal number of anion vacancies are generated. These are occupied by the electron released when the incoming metal atom is ionized. KCl becomes violet, KBr blue-green, and NaCl deep yellow. Theoretically the transitions responsible for the colour may be interpreted in terms of a particle (in this case the electron) in a box

H-centres
These are formed when the alkali metal halide is heated in the vapour of the halogen, X_2. The X_2^- anion occupies a single anion site

Crystal defects

Even a small inorganic crystal having a mass of 0.1 g is likely to have 10^{23} ions and it would be stretching reality to think that the growth of the crystal from a solution could have occurred in such a way that no impurities or imperfections were incorporated. Of course the impurities may be removed by successive purification steps, but even in the purest crystal some structural imperfections (defects) remain. The types of defects which are commonly observed can be classified as intrinsic, which are integral to the crystal and extrinsic, which are created by the insertion of foreign ions into the lattice.

Defects which maintain the same formula (stoichiometry)

In an ionic crystalline material the removal of either a cation or an anion would leave a residual charge on the crystal. If the cation and anion have only a single commonly occurring oxidation state, e.g. Na^+ or F^-, this is energetically unfavourable. However, if oppositely charged ions are removed in a pairwise fashion then the crystal stays electroneutral. The two vacancies created in the lattice are described as **Schottky defects**. The vacancies may either coexist in adjacent sites or more probably in sites which are a longer distance apart in the lattice. Simple coulombic arrangements lead to defects being more stable when adjacent—the tendency for these to separate depends on the dielectric constant. Specifically, ions in the first shell will tend to move away from the centres of the vacancies in order to reduce the repulsions between identically charged ions.

Alternatively an ion may migrate from its regular lattice site into an alternative site in the lattice which is not usually occupied. The lattice relaxes to accommodate the additional interstitial atom. For this to occur a site has to be found with approximately the same dimensions and this is more likely to occur if the cation and anion have very different sizes. A close packed structure with ions of approximately equal sizes creates only small cavities between the atoms. Although the creation of the defect involves the migration of a single ion it may also be viewed as the pairwise addition of a new interstitial atom and the creation of a vacancy. These defects are described as **Frenkel defects** and may also occur in adjacent or non-adjacent sites in the lattice. The former are described as split interstitials.

Schottky and Frenkel defects are most commonly observed in salts of the alkali metals and alkaline earth metals.

Defects which result in changes in the formula (stoichiometry)

The loss of a cation or an ion may be accommodated if the remaining ion has alternative oxidation states which can subsume the charge which is generated on its removal. For example, if a compound has an initial formula MY (M = metal, Y = anion) it can incorporate additional anions into the structure if suitable cavities can be found for them and if the metal has an oxidized form which can balance the additional negative charges introduced. For example, if the metal is Fe^{II} chemical knowledge suggests that the oxidation of some of the iron atoms to Fe^{III} will compensate for the additional negative charges resulting from the introduction of additional anions. Specifically, each interstitial oxide requires two iron atoms to be oxidized from Fe^{II} to Fe^{III}. Alternatively, some of the metal atoms could be removed to create vacancies in the lattice if a matching number of the remaining ions increase their oxidation state to compensate for the change in charge.

Table C.1 summarizes the alternative permutations observed in non-stoichiometric compounds.

Table C.1 Various formulations in which non-stoichiometric compounds are observed

Metal required to be oxidized		Metal required to be reduced	
Excess of anions, X	Metal vacancies	Excess of metal ions, M	Anion Vacancies
MY_{1+y}	$M_{1-x}Y$	$M_{1+x}Y$	MY_{1-x}
UO_{2+x}	$Fe_{1-x}O$	$Zn_{1+x}O$	WO_{3-x}

The extent of non-stoichiometry as defined by x in the general formulae above depends on a subtle interplay of structural and thermodynamic effects. Removing ions from a crystalline structure will in general result in a loss of lattice enthalpy. The pairwise nature of the electrostatic interaction between cations and anions means that this decrease will be approximately proportional to the number of ions removed providing the relaxation effects are not very great. This loss of enthalpy is compensated for by an increase in the entropic contribution to the total free energy of the system. The introduction of a vacancy into an otherwise perfect lattice increase the disorder in the system because that vacancy can be located at any one of the 10^{20} sites in the crystal. As the number of vacancies increase they can be permuted around the structure but a maximum is reached and then the entropy contribution fails to keep pace with the enthalpic contribution. Therefore, an equilibrium concentration of defects is achieved. If the salt has a high lattice energy then the loss lattice enthalpy does not permit a large number of defects. For example, NaCl is almost defect free. However, if the lattice enthalpies are smaller then the equilibrium concentration of defects is larger. The range of defects also depends on how well the lattice is able to accommodate the vacancies and interstitial by relaxation processes. For a structure involving a high defect concentration concerted structural changes may occur throughout the lattice to reduce the consequences of the defects, e.g. by crystallographic sheer distortions.

Vacancies can also be introduced into a crystal by doping it with a compound which has a different stoichiometry. For example, if $CaCl_2$ is added to NaCl each Ca^{2+} ion replaces two Na^+ ions and a cation vacancy is created to preserve electroneutrality. These are described as extrinsic vacancies. Extrinsic anion vacancies may also be produced either by replacing X^- with Y^{2-} or M^{2+} by M'^+.

Ionic conductivity

The creation of vacancies in a structure is analogous to producing empty peg holes in a solitaire board. Both have the effect of promoting movement. For an ionic crystal the movement of ions via vacancies leads to ionic conductivity. The activation energy required for the hopping of ions is generally quite large and consequently the ionic conductivity of most salts is much smaller than that of metallic electronic conductors. There are some fast ion conductors, e.g. α-AgI (above 147°C), $RbAg_4I_5$, CaO/ZrO_2, and β-alumina where the activation energy is an order of magnitude smaller because the lattices have a more open structure which enables the cations to migrate to vacancies more easily

Non-stoichiometry can have a very dramatic effect on electronic conductivity, e.g. $YBa_2Cu_3O_7$ is an insulator whereas $YBa_2Cu_3O_{6.9}$ is a superconductor with $T_c = 93$ K (see Band theory and Superconductors)

For a more detailed discussion see L. Smart and E. Moore, *Solid State Chemistry*, 2nd edn, Chapman and Hall, London, 1995 and M. T. Weller, *Inorganic Materials Chemistry*, OUP, Oxford, 1994.

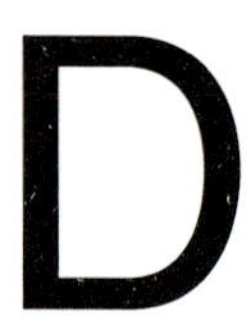

Dewar–Chatt–Duncanson bonding model

Table D.1 ν(C=C) stretching frequencies (cm^{-1}) for ethene and its complexes

	ν(C=C)
C_2H_4	1623
$K[PtCl_3(C_2H_4)]$	1584
$[(C_5H_5)Rh(C_2H_4)_2]$	1493

Table D.2 C=C bond lengths (pm) for ethene and its complexes and the C–C bond length in ethane

	C=C
C_2H_4	133.5
$K[PtCl_3(C_2H_4)]$	137.5
$[(Ph_3P)_2Ni(C_2H_4)]$	146
$[(Ph_3P)_2Pt(C_2H_4)]$	143
$[Os(CO)_4(C_2H_4)]$	149
C_2H_6	153.2

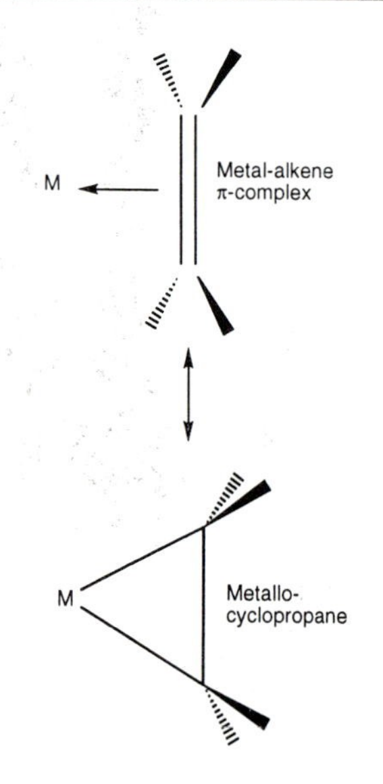

Fig. D.2 Alternative extreme representations of the bonding in metal–alkene complexes

Dewar–Chatt–Duncanson bonding model

In classical Werner coordination complexes the metal-ligand bond is described in terms of donation from a lone pair on the ligand to an empty orbital on the metal. Alkenes and alkynes have no lone pairs and form only very weak complexes with traditional Lewis Acids such as BF_3, but do form stable complexes with transition metals in low oxidation states. The Dewar–Chatt–Duncanson (DCD) model rationalizes these observations by proposing that the bonding in metal–alkene complexes depends on synergic interactions similar to those found in metal carbonyl complexes. Specifically the forward donation component involves donation from the filled π-orbital of the alkene to an empty orbital on the metal and this is supplemented by back donation from filled orbitals on the metal to empty π^* orbitals on the alkene. The effective participation of these components has symmetry constraints since the overlapping orbitals must have matching symmetry properties and these are only satisfied if the alkene bonds to the metal in the side-on (π) fashion illustrated in Fig. D.1.

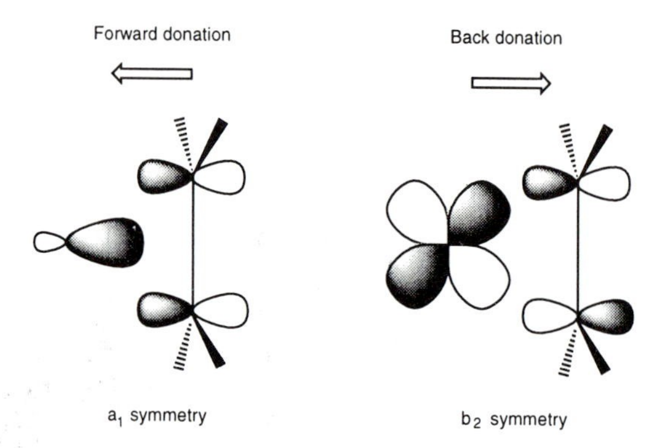

Fig. D.1 The forward and back donation components of the Dewar–Chatt–Duncanson model

The DCD model has the following structural and spectroscopic implications.

1. The alkene loses its centre of symmetry on complexation in a π-fashion and therefore the ν(C=C) stretching frequency which is infrared inactive in the parent alkene becomes infrared active in the complex.
2. The donation of electron density from the filled alkene π-orbital and acceptance of electron density from the metal to the alkene π^*–orbitals both reduce the strength of the C–C bond and this is manifested in a longer C–C bond distance in the complex and a lower ν(C=C) stretching frequency and force constant. The data given in Tables D.1 and D.2 illustrate these consequences.
3. The forward and back donation components change the hybridization at the carbon atoms from sp^2 to sp^3 and this has the effect of changing the H–C–H and C=C–H bond angles and the extent to which the hydrogen atoms bend back away from the metal.

In valence bond terms the bonding may be described in terms of the resonance forms shown in Fig. D.2, with the metal cyclopropane structure predominating in complexes where the extent of back donation is extensive.

The relative importance of the forward and back donation components depends on a range of factors and extensive structural studies have resulted in the following generalizations.

The metal oxidation state, ligands coordinated to the metal and substituent effects on the alkenes have a greater influence on the back donation component than the forward donation component

Electron withdrawing effects on the alkene enhance the degree of back donation and therefore lead to alkene complexes where the C–C bond lengths are longer and the metal–carbon bond lengths are shorter. Fig. D.3 gives some specific examples of these effects.

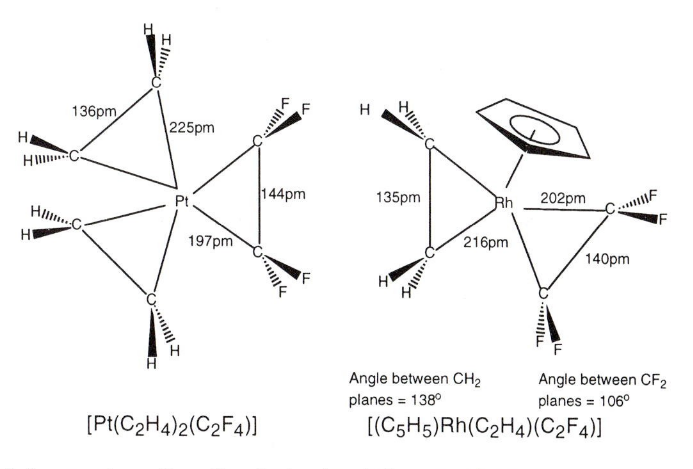

Fig. D.3 A comparison of bond lengths in related ethene and fluoroethene complexes. The fluorine atom is strongly electron-withdrawing

Lowering the metal oxidation state or introducing more electron donating substituents on the metal enhance the back donation effects. See Fig. D.4, for example, which illustrates the effect of simultaneously lowering the oxidation state and introducing more electron donating ligands.

The coordination of an alkene also influences its reactivity towards nucleophiles and electrophiles. For example, coordination of alkenes to positively charged metal ligand fragments which are not capable of back donating extensively causes the alkene to become more susceptible to nucleophilic attack.

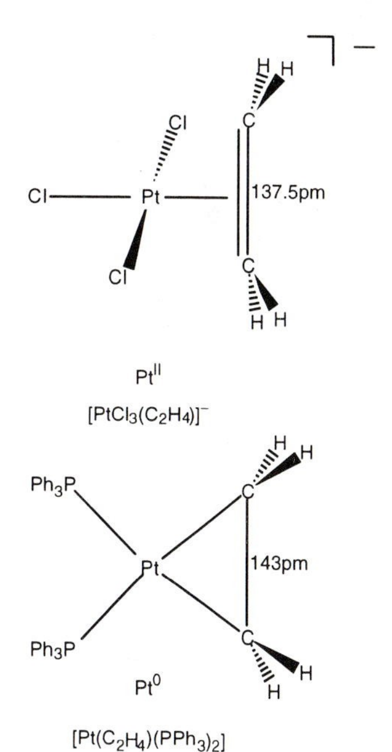

Fig. D.4 Bond lengths in platinum(II) and platinum(0) ethene complexes

Rotational barriers

The alkene is capable of rotating about the vector from the metal to the centre of the alkene, however the activation energy for the process is very dependent on the back donating abilities of the metal-ligand fragment and steric effects. If the metal-ligand has a pair of degenerate d_{xz} and d_{yz} orbitals then the rotational barrier is very low (< 40 kJ mol^{-1}). However, if they have very different energies then one of this pair is much more able to back donate than the second and therefore their is a preferred ground state conformation and the rotational barrier is large (> 80 kJ mol^{-1}). Fig. D.5 illustrates some specific examples of complexes with small and large rotational barriers.

The DCD model is applicable to a wide range of other ligands which have a π–donor and a π*–acceptor orbital and some relevant examples are illustrated in Fig. D.6.

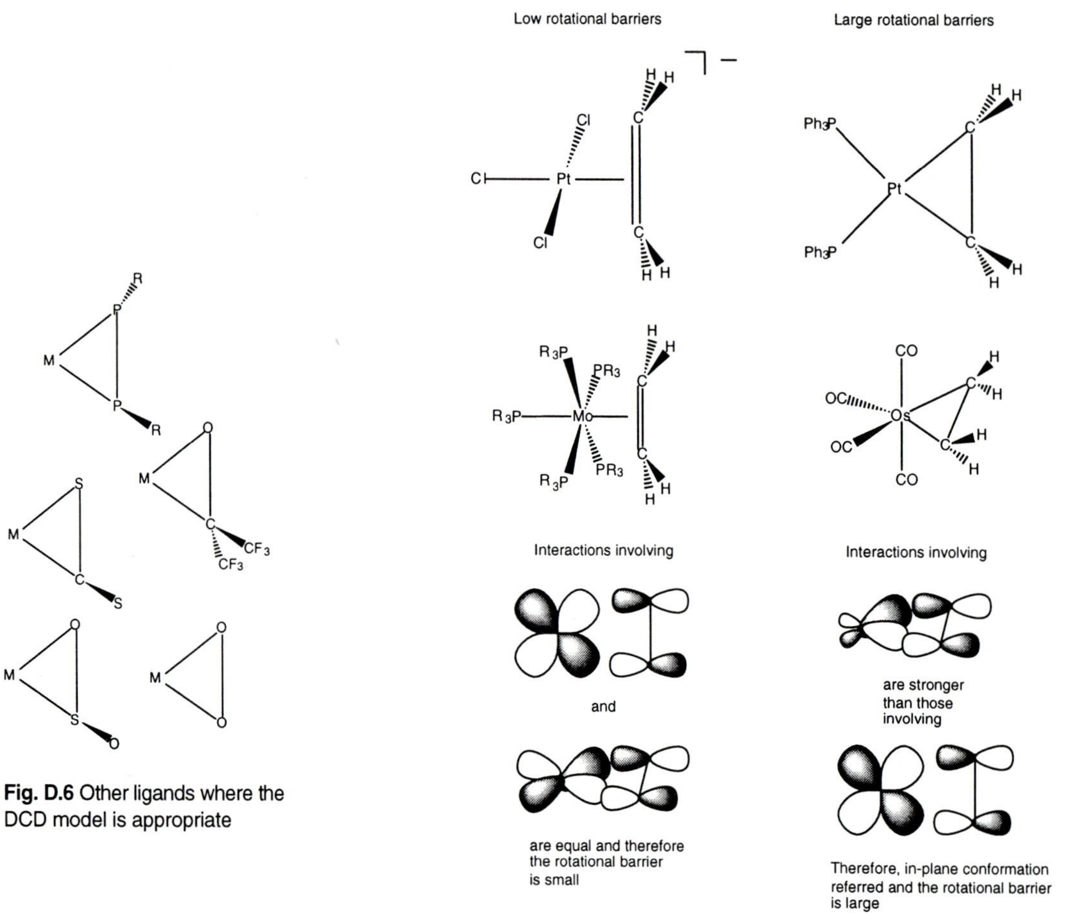

Fig. D.6 Other ligands where the DCD model is appropriate

Fig. D.5 Relationships between orbital interactions and rotational barriers in metal–alkene complexes

It may also be developed to account for the structural properties of polyalkene complexes. For example, coordination of butadiene to a transition metal complex leads to a equalization of the C–C bond lengths. This may be interpreted in terms of the nodal properties of the primary donor and acceptor orbitals of the diene shown in Fig. D.7.

If the ligand has cylindrical symmetry, e.g. alkynes, there are two π and π* orbitals to be considered. The organic substituents on the alkyne bend away from the metal in much the same way as that described above for alkenes and the forward and back donation components in the metal–C—C plane, illustrated in Fig. D.8 involve very similar bonding interactions to those described above. There is an additional filled π orbital in a perpendicular plane which is capable of donating electron density to an empty orbital on the metal, however its π* component has δ symmetry with respect to the metal and therefore the overlap is not sufficiently large to lead to a big back donation component.

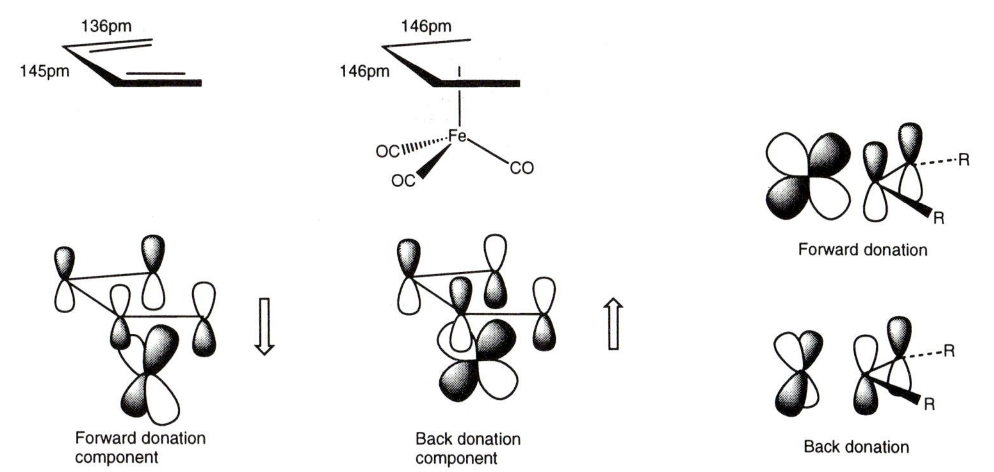

Fig. D.7 Forward and back donation components in metal–butadiene complexes

R
R
Forward donation

R
R
Back donation

Fig. D.8 The orbitals of a metal atom and those of an alkene participating in forward and back donation

Therefore, alkynes are able to function as either a two-electron ligand in metal complexes or a four-electron ligand depending on whether the metal has suitable empty orbitals available. The examples shown in Fig. D.9 illustrate these alternative bonding modes for a pair of molybdenum complexes.

M. Bochmann, *Organometallics 2: Complexes with Transition Metal–Carbon π-bonds*, OUP, Oxford, 1994, provides a detailed account of this topic

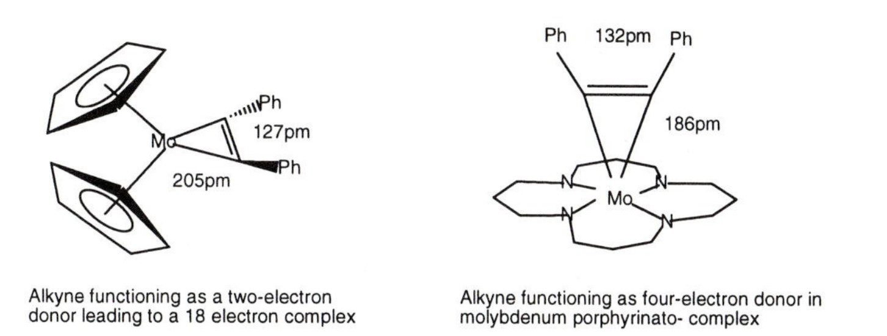

Fig. D.9 Alkyne complexes where the ligand can function either as a two- or four-electron donor

Group theory

It would be helpful to read the **Symmetry** section prior to reading this one

Group theory

The classification of molecules according to their symmetry characteristics is described in *symmetry* on page 76. Following the procedures described in that section it is possible to unambiguously assign a point group to any molecule and designate it a Schönflies symbol. This mode of classification of molecules is able to establish whether a molecule may have a dipole moment or may be optically active, but in itself does not provide any information about the electronic structure and spectroscopic properties of the molecule. The inter-relationships between the symmetry properties and these observable properties are defined using a mathematical formalism described as Group Theory. The mathematical basis of Group Theory is beyond the scope of this book, but the underlying principles may be appreciated. The properties of a molecule of interest to a chemist such as its total energy, the number of infrared vibrational modes, the number of electronic transitions, etc. must be invariant to the symmetry operations. It would clearly be a nonsense if the energy of a molecule were to change on the application of a symmetry operation which merely permutes the positions of identical atoms. Secondly, the solution of a Schrödinger equation for the hydrogen atom leads to sets of degenerate orbitals, which are crucial to chemists, because their number, shape and energies not only determine the form of the Periodic Table but also the number and type of bonds in molecules. An isolated atom has spherical symmetry, but the environment of an atom within a molecule is less symmetrical and therefore the degeneracies associated with the free atom are partially or completely lost in a molecule. For an atom located at the origin of the point group, where all the symmetry operations intersect, the loss of degeneracy may be directly related to the point group of the molecule. Table G.1 traces this loss of degeneracy for p and d orbitals in the I_h (icosahedral), O_h (octahedral), D_{4h} (square planar), and C_{2v} (angular) point groups, where the principal axis is aligned along z.

Table G.1 Loss of degeneracies in specific point groups. Sets of degenerate orbitals are shown in boxes

Spherical	Icosahedral I_h	Octahedral O_h	Square planar D_{4h}	Angular C_{2v}
s	s	s	s	s
p	p_x, p_y, p_z	p_x, p_y, p_z	p_x, p_y	p_x
			p_z	p_y
				p_z
d	$d_{z^2}, d_{x^2-y^2}, d_{xy}, d_{xz}, d_{yz}$	d_{xy}, d_{xz}, d_{yz}	d_{xz}, d_{yz}	d_{z^2}
		$d_{z^2}, d_{x^2-y^2}$	d_{z^2}	$d_{x^2-y^2}$
			$d_{x^2-y^2}$	d_{xy}
			d_{xy}	d_{xz}
				d_{yz}

It is noteworthy that since the spherical s orbital is not degenerate it remains unaffected by the descent in symmetry across the Table. The triply degenerate p orbitals retain their triple degeneracy in the icosahedral and octahedral point groups, but in the D_{4h} point group their degeneracy is lost.

The d orbitals remain degenerate in the icosahedral point group, I_h, but split into two sets in the octahedral point group, O_h, and four sets in the planar point group D_{4h}. Clearly as the environment around the atom becomes less spherical then the degeneracy is increasingly removed.

In a octahedral complex the three p orbitals point directly at the ligands and they are indistinguishable because successive 120° rotations (C_3 and C_3^2) along the three-fold axis interchanges them (see diagrams in the margin).

A complex with a square planar coordination geometry belongs to the point group D_{4h}. Two of the p orbitals (p_x and p_y) point towards the ligands whereas the p_z points along the four fold axis and away from the ligands. This qualitative difference may be expressed in a more precise way mathematically by recognising that the symmetry operations of the D_{4h} point group may interchange p_x and p_y, but do not interchange p_z with either p_x or p_y.

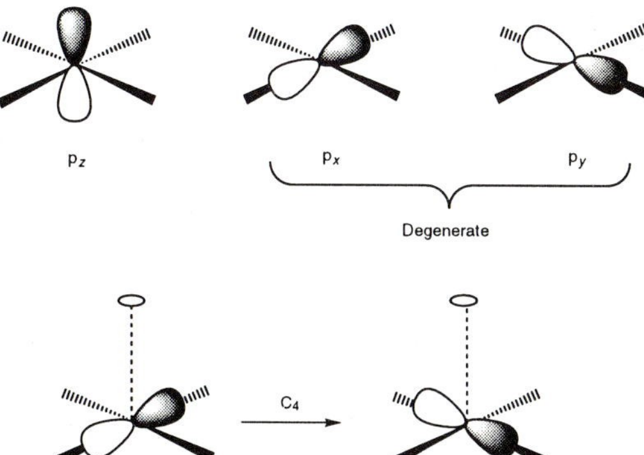

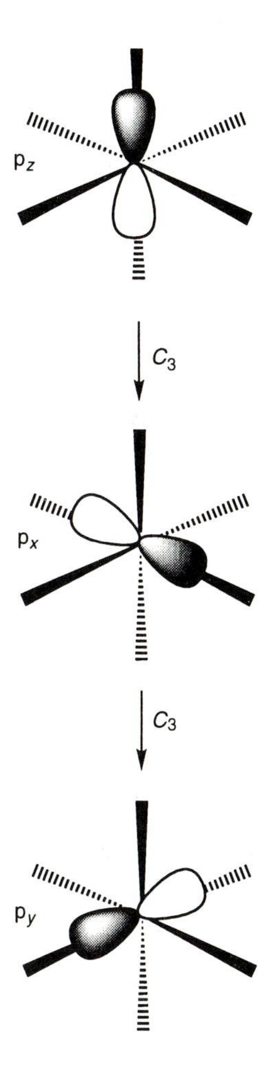

The C_3 axis is perpendicular to the page

For example, the application of a 90° rotation (C_4) along the z axis leaves p_z unaffected and interchanges p_x and p_y. Algebraically this may be represented by the following relationships:

$$C_4 \times p_z = p_z$$
$$C_4 \times p_x = p_y$$
$$C_4 \times p_y = p_x$$

Alternatively the C_4 operation may be represented by a matrix as follows:

$$\begin{bmatrix} 1 & 0 & 0 \\ 0 & 0 & 1 \\ 0 & 1 & 0 \end{bmatrix} \begin{bmatrix} p_z \\ p_x \\ p_y \end{bmatrix} = \begin{bmatrix} p_z \\ p_y \\ p_x \end{bmatrix}$$

The corresponding matrix for the 120° rotation around the three-fold axis of rotation of p orbitals in an octahedral complex is:

$$\begin{bmatrix} 0 & 1 & 0 \\ 0 & 0 & 1 \\ 0 & 1 & 0 \end{bmatrix} \begin{bmatrix} p_z \\ p_y \\ p_x \end{bmatrix} = \begin{bmatrix} p_x \\ p_z \\ p_y \end{bmatrix}$$

Recall that matrix multiplication involves the successive multiplication of rows in the first matrix by the columns of the second matrix

$$\begin{bmatrix} a & b \\ c & d \end{bmatrix}\begin{bmatrix} x \\ y \end{bmatrix} = \begin{bmatrix} ax+by \\ cx+dy \end{bmatrix}$$

This matrix clearly documents the permutation of these three orbitals which retain their degeneracy. The degeneracy of the representation governs the dimensions of the matrices used to describe the symmetry operations. The degeneracy of the orbitals is transparent from the matrices since the orbitals which are degenerate have off-diagonal elements. The matrix for the 90° rotation in a D_{4h} square planar complex may be blocked a way that emphasizes their degeneracies as follows.

$$\begin{matrix} [1] \\ \begin{bmatrix} 0 & 1 \\ 1 & 0 \end{bmatrix} \end{matrix} \begin{matrix} [p_z] \\ \begin{bmatrix} p_x \\ p_y \end{bmatrix} \end{matrix} = \begin{matrix} [p_z] \\ \begin{bmatrix} p_y \\ p_x \end{bmatrix} \end{matrix}$$

More generally for a rotation of θ around the *z* axis the matrix takes the form:

$$\begin{bmatrix} \cos\theta & \sin\theta & 0 \\ -\sin\theta & \cos\theta & 0 \\ 0 & 0 & 1 \end{bmatrix}\begin{bmatrix} p_x \\ p_y \\ p_z \end{bmatrix}$$

The trace of the matrix is 1 + 2cosθ

The representation of symmetry operations by matrices may appear initially as a cumbersome, but mathematically precise, way of recording the effect of the symmetry operations. Fortunately, it is not necessary to retain all the information contained within the matrix. Specifically, the diagonal elements of the matrix contain all the information required to obtain the most significant conclusions. The sum of the diagonal elements of the matrix—called the trace or character of the matrix—contains the essential symmetry information.

For a class of symmetry operations the traces of the matrices are identical. Therefore, once the trace of a particular symmetry operation has been established then it is not necessary to consider the other operations in the point group belonging to the same class. For example, in the octahedron there are $8C_3$ operations in a class, but the traces of the matrices representing these operations are identical and equal to that given in the matrix above, i.e. 0.

For a definition of *class* see **Symmetry**

Since the labels s, p, d, and f no longer adequately describe the symmetry properties of the orbitals in non-spherical situations it is necessary to introduce a different notation. The nomenclature proposed by R. S. Mulliken, who was awarded the Nobel Prize for chemistry for his pioneering work on molecular orbital theory, is used universally.

All singly degenerate representations are designated either A or B; doubly degenerate representations are designated E and triply degenerate representations T.

One dimensional representations which are symmetric with respect to rotation by 360°/n about the principal C_n *axis are designated A, whilst those which are anti-symmetric are represented by B.*

Subscripts are attached to A or B to indicate whether they are symmetric or antisymmetric with respect to a C_2 *axis perpendicular to the principal axis or if such an axis is absent the vertical mirror plane.*

The subscripts g and u are added to representations which are symmetric or antisymmetric with respect to the inversion operation in point groups which have a centre of symmetry.

Primes and double primes are used to indicate whether the representations are symmetric or antisymmetric with respect to a horizontal mirror plane in the point group.

Table G.2 illustrates the application of the Mulliken nomenclature to the s, p, and d valence orbitals of an atom located at the origin of a number of different point groups.

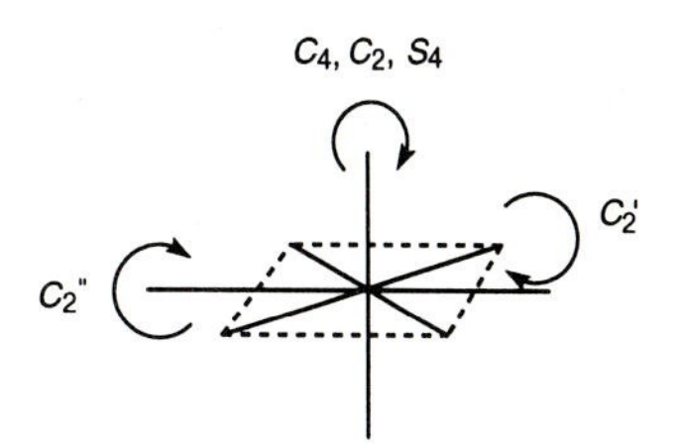

Fig. G.1 The axes of symmetry for a square planar complex belonging to the D_{4h} point group

Table G.2 Mulliken notation for the s, p, and d orbitals in a range of point groups

Spherical	Icosahedral, I_h	Octahedral, O_h	Square planar, D_{4h}	Angular, C_{2v}
s	a_{1g}	a_{1g}	a_{1g}	a_1
p_x	} t_{1u}	} t_{1u}	} e_u	b_1
p_y				b_2
p_z			a_{2u}	a_1
d_{z^2}	} h_g	} e_g	a_{1g}	a_1
$d_{x^2-y^2}$			b_{1g}	a_1
d_{xy}		} t_{2g}	b_{2g}	a_2
d_{xz}			} e_g	b_1
d_{yz}				b_2

It is clear that for each point group there are a set of distinctive representations which represent alternative combinations of the traces of the symmetry transformation matrices. Some of them match the transformation properties of orbital functions on an atom located at the origin. This information is summarized in a *character table.* The ability to manipulate the information contained within a character table is essential for solving a wide range of spectroscopic and bonding problems in chemistry. A typical character table, the one for the D_{4h} point group, is illustrated in Table G.3.

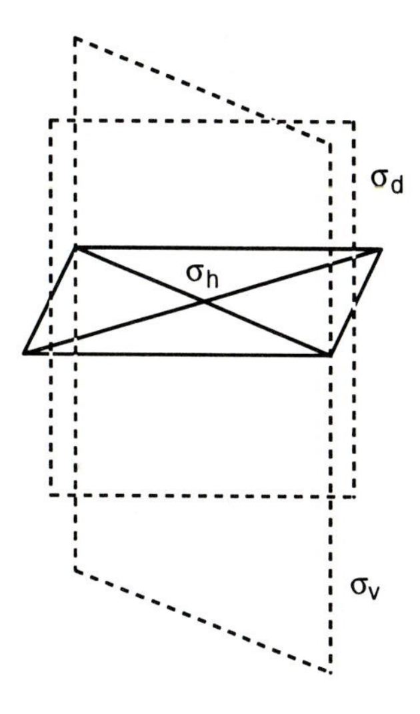

Fig. G.2 The planes of symmetry of the D_{4h} point group

The Schönflies symbol for the character table is indicated in the top left hand corner and below it in a column are the Mulliken symbols which were introduced above. Although the Mulliken symbols are given in upper case letters in the character table, there is a convention that lower case symbols are used for orbitals and upper case symbols for electronic states. The D_{4h} point group has a centre of symmetry and consequently there are g and u subscripts indicating whether the function is symmetric or antisymmetric to the centre of inversion. The top row of the table lists the classes of symmetry operations. The total number of symmetry operations in the point group is the sum of the numbers of operations in each class. In this specific case there are a total of 16 symmetry operations in 10 classes.

Below the symmetry operations are given the characters (traces) of the relevant matrices for the symmetry operations. The group theoretical analysis proves that the number of rows in the D_{4h} table is equal to the number of classes of symmetry operations. Such rows are described as irreducible representations of the point group. They are described as irreducible because they represent the simplest representations of the alternative transformation properties of functions which arise from combining the operations associated with the distinct classes. The first row of the table represents a function which is totally symmetric to all the symmetry operations of the point group. Therefore every entry has +1 associated with it. This for obvious reasons is described as the *totally symmetric representation.*

At the right-hand side of the table are a set of algebraic functions which mirror the transformation properties of the irreducible representation. For example, for the D_{4h} point group the p_z orbital, which transforms as z, is symmetric to C_4, but antisymmetric to the C_2 axes perpendicular to it. Therefore, the entries are +1, −1, and −1 under the headings $2C_4$, $2C_2'$, and $2C_2''$.

Table G.3 The character table for the D_{4h} point group

D_{4h}	E	$2C_4$	C_2	$2C_2'$	$2C_2''$	i	$2S_4$	σ_h	$2\sigma_v$	$2\sigma_d$		
A_{1g}	1	1	1	1	1	1	1	1	1	1		x^2+y^2, z^2
A_{2g}	1	1	1	−1	−1	1	1	1	−1	−1	R_z	
B_{1g}	1	−1	1	1	−1	1	−1	1	1	−1		x^2-y^2
B_{2g}	1	−1	1	−1	1	1	−1	1	−1	1		xy
E_g	2	0	−2	0	0	2	0	−2	0	0	(R_x, R_y)	(xy, yz)
A_{1u}	1	1	1	1	1	−1	−1	−1	−1	−1		
A_{2u}	1	1	1	−1	−1	−1	−1	−1	1	1	z	
B_{1u}	1	−1	1	1	−1	−1	1	−1	−1	1		
B_{2u}	1	−1	1	−1	1	−1	1	−1	1	−1		
E_u	2	0	−2	0	0	−2	0	2	0	0	(x, y)	

The characters of the irreducible representations in tables such as Table G.3 have the following properties.

The sum of the squares of the dimensions of the irreducible representations of a group is equal to the order of the group.

Matrices for the identity operation, E, for singly, doubly, and triply degenerate representations

$$[1]$$

$$\begin{bmatrix} 1 & 0 \\ 0 & 1 \end{bmatrix}$$

and

$$\begin{bmatrix} 1 & 0 & 0 \\ 0 & 1 & 0 \\ 0 & 0 & 1 \end{bmatrix}$$

As an example of this general rule, the irreducible representations of the point group have degeneracies equal to the characters shown in the E column of the character table (Table G.3): 1, 1, 1, 1, 2, 1, 1, 1, 1, and 2 respectively and the sum of their squares is 16—the order of the D_{4h} point group.

For the singly degenerate representations in the D_{4h} point group this means that the characters must all be either +1 or −1. For the doubly degenerate representations this is satisfied by having some of the characters equal to 2 and others equal to zero. For triply degenerate representations the characters can be ±3, ±2, ±1, or 0. The character of the representation associated with the E operation must correspond to the degeneracy of the operation since, if the operation consists of doing nothing and the character is the sum of the diagonal elements, the relevant matrices for singly, doubly, and triply degenerate representations are shown in the margin, i.e. they are simply unit matrices.

The vectors whose components are the characters of two irreducible representations are orthogonal.

The character tables have been introduced above as a descent in symmetry from spherical and it will be recalled that the spherical harmonic functions are orthogonal and therefore this orthogonality property is retained in the irreducible representations.

The d orbitals in a spherical atom are degenerate, but in a molecule with D_{4h} symmetry they fall into four groups. If we add the characters, Γ, of the irreducible representations of the five d orbitals for the D_{4h} point group we obtain the following result:

	E	$2C_4$	C_2	$2C_2'$	$2C_2''$	i	$2S_4$	σ_h	$2\sigma_v$	$2\sigma_d$	
$\Gamma(a_{1g})$	1	1	1	1	1	1	1	1	1	1	d_{z^2}
$\Gamma(b_{1g})$	1	–1	1	1	–1	1	–1	1	1	–1	$d_{x^2-y^2}$
$\Gamma(b_{2g})$	1	–1	1	–1	1	1	–1	1	–1	1	d_{xy}
$\Gamma(e_g)$	2	0	–2	0	0	2	0	–2	0	0	d_{xz}, d_{yz}
Γ_R	5	–1	1	1	1	5	–1	1	1	1	

The final line is described as a reducible representation since it represents the sum of the four irreducible representations. Reducible representations may be generated not only for atomic orbitals centred on the central atom, but also for linear combinations of orbitals on the peripheral atoms of the molecule. Displacement vectors of the atoms which are important for the analysis of the number of infrared and Raman active modes, etc., may also form the basis of reducible representations. Therefore, the ability to unravel the irreducible components of a reducible representation is an essential skill which has to be gained in the study of Group Theory. There is a fundamental formula which tests a reducible representation in order to find out whether it contains a particular irreducible representation. It may be established whether an irreducible representation is contained within a reducible one by treating them as vectors and taking a dot product divided by the order using the equation:

The dot product of the two vectors:
$ax + by + cz + de$
$a'x + b'y + c'z + d'e$
is:
$a \times a' + b \times b' + c \times c' + d \times d'$
The vectors are orthogonal if this dot product is zero

$$a_i = \frac{1}{h}\sum_g g_R \chi(R)\chi_i(R) \qquad \text{(G.1)}$$

where h is the order of the group—the total number of symmetry operations, g_R is the number of symmetry operations in the class, $\chi_i(R)$ is the character of the irreducible representation which is being tested as a possible contributor to the reducible representation, and $\chi(R)$ is the character of the reducible representation.

For example, to confirm that the irreducible representation for $d_{x^2-y^2}$ (b_{2g}) contributed to Γ_R above the formula works as follows:

	E	$2C_4$	C_2	$2C_2'$	$2C_2''$	i	$2S_4$	σ_h	$2\sigma_v$	$2\sigma_d$
χ_R	5	–1	1	1	1	5	–1	1	1	1
$\chi_{(i)}b_{2g}$	1	–1	1	1	–1	1	–1	1	1	–1

$h = 16$

$$a(b_{2g}) = \frac{1}{16}[1.5.1 + 2.-1.-1 + 1.1.1 + 2.1.1 + 2.1.-1 + 1.5.1 + 2.-1.-1 + 1.1.1 + 2.1.1 + 2.1.-1] = \frac{16}{16} = 1$$

However, for p_z which belongs to the irreducible representation a_{2u} the formula gives:

$$a(a_{2u}) = \frac{1}{16}[1.5.1 + 2.-1.1 + 1.1.1 + 2.1.-1 + 2.1.-1 + 1.5.-1 + 2.-1.-1 + 1.1.-1 + 2.1.1 + 2.1.1] = 0$$

This confirms that the reducible representation does not contain a contribution from a_{2u}.

The following sections illustrate how the character table and equation (G.1) can be used to solve a range of bonding and spectroscopic problems.

Derivation of qualitative molecular orbital diagrams

A tetrahedral EH_4 molecule (E = a main group atom) belongs to the T_d point group and the relevant character table is given below.

T_d	E	$8C_3$	$3C_2$	$6S_4$	$6\sigma_d$	
A_1	1	1	1	1	1	$x^2 + y^2 + z^2$ (s)
A_2	1	1	1	–1	–1	
E	2	–1	2	0	0	$2z^2 - x^2 - y^2, x^2 - y^2$
T_1	3	0	–1	1	–1	R_x, R_y, R_z
T_2	3	0	–1	–1	1	(x, y, z) (xz, yz, xy)

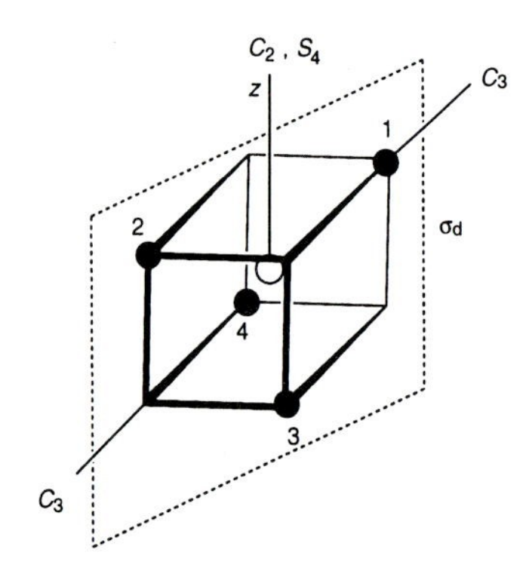

Fig. G.3 The locations of the symmetry operations of the T_d point group

The location of specific examples of the classes of symmetry operation are illustrated in Fig. G.3.

The transformation properties of the atomic orbitals of the central atom may be obtained directly from the character table. They are *ns*—a_1 $(x^2 + y^2 + z^2)$ and *np*—t_2 (x, y, z).

The symmetry transformation properties of the hydrogen 1s orbitals may be represented by the matrices shown in the margin opposite which act on the vector containing the atom numbers and therefore indicate how the 1s orbitals move as a result of the symmetry operations indicated in Fig. G.3. The traces of the matrices are indicated below each one.

These traces correspond to the characters of a reducible representation, Γ_R, which encompasses the symmetry transformation properties of the four hydrogen 1s orbitals.

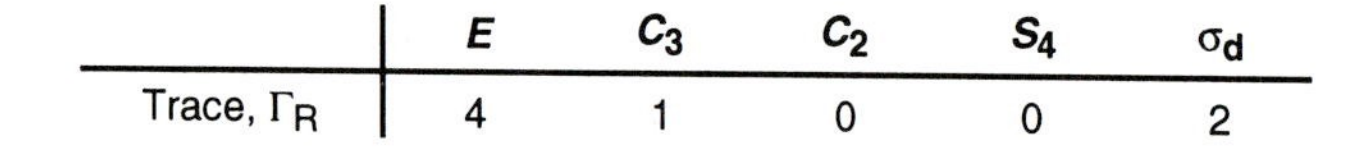

	E	C_3	C_2	S_4	σ_d
Trace, Γ_R	4	1	0	0	2

Conventions for group theory symbols

Lower case symbols are used for:
(a) normal vibrational modes and
(b) one-electron wavefunctions.
Capitals are used for:
(a) vibrational state wavefunctions,
(b) electronic state wavefunctions, and
(c) vibronic state wavefuctions.

The characters of the reducible representation may have been obtained more simply by asking the following question for each operation: *How many hydrogen atoms remain in the same position after the execution of the symmetry operation?* This number corresponds to the trace of the matrix, because only atoms which remain in the same position contribute to the diagonal elements of the matrix.

The irreducible representations, Γ_i, which contribute to Γ_R may be obtained by applying the formula given in equation (G.1).

$$a(a_1) = \frac{1}{24}[1.4.1 + 8.1.1 + 3.0.1 + 6.0.1 + 6.2.1] = \frac{24}{24} = 1$$

$$a(a_2) = \frac{1}{24}[1.4.1 + 8.1.1 + 3.0.1 + 6.0.{-1} + 6.2.{-1}] = \frac{0}{24} = 0$$

$$a(e) = \frac{1}{24}[1.4.2 + 8.1.{-1} + 3.0.2 + 6.0.0 + 6.2.0] = \frac{0}{24} = 0$$

$$a(t_1) = \frac{1}{24}\,[1.4.3 + 8.1.0 + 3.0.{-1} + 6.0.1 + 6.2.{-1}] = \frac{0}{24} = 0$$

$$a(t_2) = \frac{1}{24}\,[1.4.3 + 8.1.0 + 3.0.{-1} + 6.0.{-1} + 6.2.1] = \frac{24}{24} = 1$$

$$\Gamma_\sigma = a_1 + t_2$$

Therefore, the hydrogen 1s orbitals transform as $a_1 + t_2$. It is noteworthy that the degeneracies of these representations add up to four which corresponds to the number of hydrogen 1s orbitals which form the basis of the reducible representation. This must always be the case and it provides a useful check for the procedure. Also, the application of the formula must always result in integer solutions—a fractional result is a sure indication that there is something wrong at an earlier stage, usually in the application of the symmetry operations.

The symmetry analysis developed above forms the basis for proposing a qualitative molecular orbital diagram for a tetrahedral EH_4 molecule. Such a diagram is shown in Fig. G.4. On the left-hand side the atomic orbitals of the central main group atom (*n*s and *n*p) are shown and the symmetries of the linear combinations of the hydrogen 1s orbitals are shown on the right-hand side. The latter are placed lower than the former on the assumption that H is more electronegative than E.

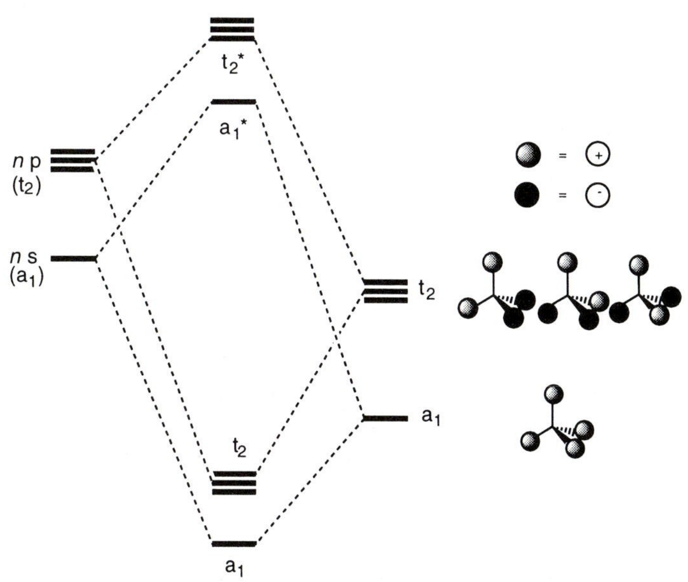

Fig. G.4 A qualitative molecular orbital diagram for a tetrahedral EH_4 molecule

The molecular orbital diagram is constructed on the basis that only orbitals with matching symmetries combine to form an in-phase (bonding) and an out-of-phase (antibonding) combination. Consequently there is a total of four bonding (a_1+t_2) and four antibonding (a_1^*+t_2^*) molecular orbitals. Therefore, the resulting molecule can accommodate a total of 8 valence

E matrix:

$$\begin{bmatrix} 1 & 0 & 0 & 0 \\ 0 & 1 & 0 & 0 \\ 0 & 0 & 1 & 0 \\ 0 & 0 & 0 & 1 \end{bmatrix}$$

Trace (*E*) = 4

C_3 matrix:

$$\begin{bmatrix} 1 & 0 & 0 & 0 \\ 0 & 0 & 0 & 1 \\ 0 & 1 & 0 & 0 \\ 0 & 0 & 1 & 0 \end{bmatrix}$$

Trace (C_3) = 1

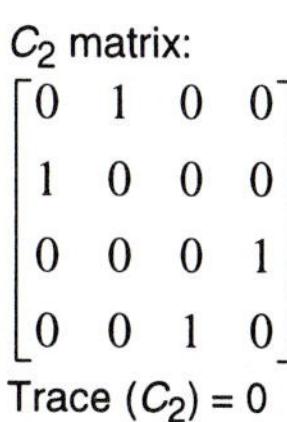

C_2 matrix:

$$\begin{bmatrix} 0 & 1 & 0 & 0 \\ 1 & 0 & 0 & 0 \\ 0 & 0 & 0 & 1 \\ 0 & 0 & 1 & 0 \end{bmatrix}$$

Trace (C_2) = 0

S_4 matrix:

$$\begin{bmatrix} 0 & 0 & 1 & 0 \\ 0 & 0 & 0 & 1 \\ 0 & 1 & 0 & 0 \\ 1 & 0 & 0 & 0 \end{bmatrix}$$

Trace (S_4) = 0

σ_d matrix:

$$\begin{bmatrix} 1 & 0 & 0 & 0 \\ 0 & 1 & 0 & 0 \\ 0 & 0 & 0 & 1 \\ 0 & 0 & 1 & 0 \end{bmatrix}$$

Trace (σ_d) = 2

electrons in the bonding a_1 and t_2 molecular orbitals and the molecular orbital diagram provides a specific illustration of the effective atomic number (EAN) or octet rule for main group molecules. Molecules and ions such as CH_4, NH_4^+, BH_4^-, and SiH_4 all of which have eight valence electrons provide specific examples of the relevance of such an analysis.

The procedure may also be illustrated for an octahedral complex and the relevant character table (O_h) is shown below.

O_h	E	$8C_3$	$6C_2$	$6C_4$	$3C_2(\equiv C_4^2)$	i	$6S_4$	$8S_6$	$3\sigma_h$	$6\sigma_d$		
A_{1g}	1	1	1	1	1	1	1	1	1	1	$x^2 + y^2 + z^2$ (s)	
A_{2g}	1	1	–1	–1	1	1	–1	1	1	–1		
E_g	2	–1	0	0	2	2	0	–1	2	0		$2z^2 - x^2 - y^2$, $x^2 - y^2$
T_{1g}	3	0	–1	1	–1	3	1	0	–1	–1	R_x, R_y, R_z	
T_{2g}	3	0	1	–1	–1	3	–1	0	–1	1		xz, yz, xy
A_{1u}	1	1	1	1	1	–1	–1	–1	–1	–1		
A_{2u}	1	1	–1	–1	1	–1	1	–1	–1	1		
E_u	2	–1	0	0	2	–2	0	1	–2	0		
T_{1u}	3	0	–1	1	–1	–3	–1	0	1	1	x, y, z	
T_{2u}	3	0	1	–1	–1	–3	1	0	1	–1		

The locations of the specific symmetry operations are illustrated in Fig. G.5

In a molecule MH_6 (where M is a transition metal) the transformation properties of the nd, (n + 1)s and (n + 1)p valence orbitals are taken directly from the character table: a_{1g} (s), t_{1u} (p), and $e_g + t_{2g}$ (d).

The linear combinations of the hydrogen 1s orbitals may be derived by assessing how many of them remain in position after the execution of the relevant symmetry operation:

	E	$8C_3$	$6C_2$	$6C_4$	$3C_2$	i	$6S_4$	$8S_6$	$3\sigma_h$	$6\sigma_d$
Γ_R	6	0	0	2	2	0	0	0	4	2

Application of the test formula given in equation (1) results in the following non-zero solutions:

$$a(a_{1g}) = \frac{1}{48}\,[1.6.1 + 8.0.1 + 6.0.1 + 6.2.1 + 3.2.1 + 1.0.1 + 6.0.1 + 8.0.1 + 3.4.1 + 6.2.1] = \frac{48}{48} = 1$$

$$a(e_g) = \frac{1}{48}\,[1.6.2 + 8.0.{-1} + 6.0.0 + 6.2.0 + 3.2.2 + 1.0.2 + 6.0.0 + 8.0.{-1} + 3.4.2 + 6.2.0] = \frac{48}{48} = 1$$

$$a(t_{1u}) = \frac{1}{48}\,[1.6.3 + 8.0.0 + 6.0.{-1} + 6.2.1 + 3.2.{-1} + 1.0.{-3} + 6.0.{-1} + 8.0.0 + 3.4.1 + 6.2.1] = \frac{48}{48} = 1$$

$$\Gamma_\sigma = a_{1g} + e_g + t_{1u}$$

Therefore, the hydrogen 1s orbitals transform as $a_{1g} + e_g + t_{1u}$, and the results form the basis of the qualitative molecular orbital diagram illustrated in Fig. G.6. The matching of orbitals with identical symmetries is an essential aspect of constructing the molecular orbital diagram. It is noteworthy that the t_{2g} metal orbitals do not find a symmetry match amongst the hydrogen linear combinations and remain localized on the metal as non-bonding orbitals. Once again the hydrogen 1s orbitals are placed below the metal valence orbitals in order to emphasize the electronegativity difference. The occupation of all the bonding and non-bonding orbitals in an octahedral complex, e.g. $[FeH_6]^{4-}$ leads to a total valence electron count of 18, i.e. it corresponds to that predicted by the Effective Atomic Number (EAN) Rule.

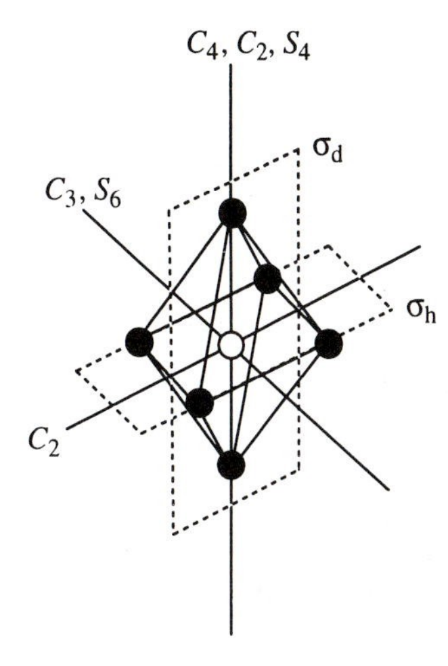

Fig. G.5 Locations of symmetry operations of the O_h point group

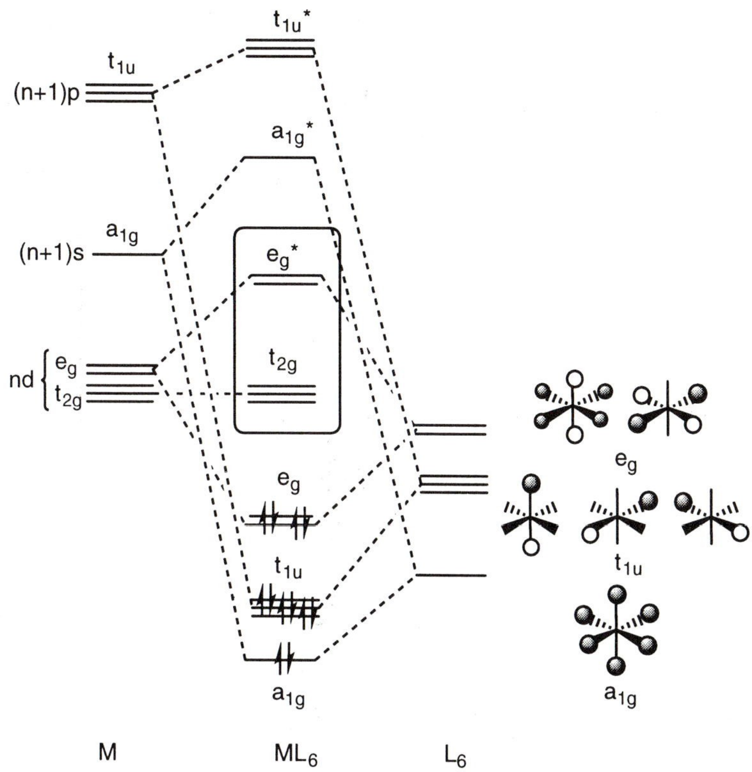

Fig. G.6 A qualitative molecular orbital diagram for an octahedral molecule. The orbitals which are localized mainly on the metal are shown in the box

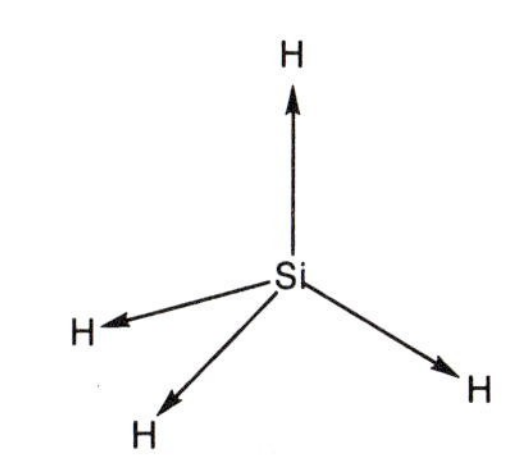

Fig. G.7 The tetrahedral SiH_4 molecule

Symmetry analysis of stretching modes

The analysis of the symmetries of stretching modes in molecules is invaluable for analysing the structures of molecules from infrared and Raman spectroscopic data. The mode of analysis is very similar to that developed above. For example, the tetrahedral molecule shown in Fig. G.7 has the four vectors representing the displacements associated with the stretching modes. In the T_d point group the vectors transform as $a_1 + t_2$.

The mode of analysis may be extended to analyse the number of stretching modes in the carbonyl region for an octahedral metal–carbonyl complex by considering the transformation properties of the CO vectors, see Fig. G.8 for example. These vectors have identical symmetry transformation properties to the MH_6 problem dealt with above, i.e. $a_{1g} + e_g + t_{1u}$.

The modes are infra-red active if they have the same transformation properties as x, y, and z and Raman active if they transform in the same way as x^2, y^2, z^2, xz, yz, and/or xy in the character table.

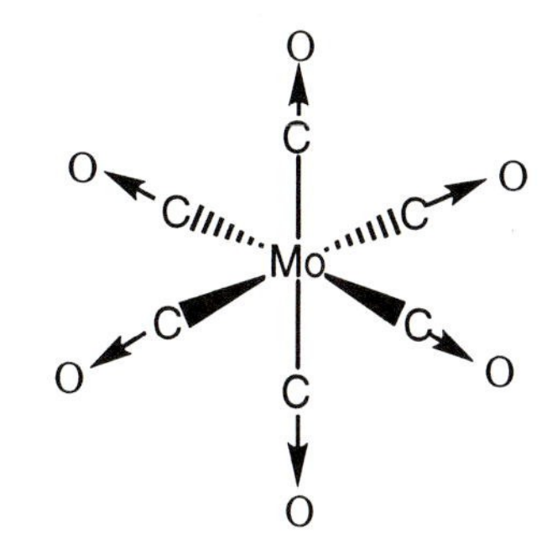

Fig. G.8 The octahedral $Mo(CO)_6$ molecule

Direct products

Many spectroscopic and bonding problems involve integrals of the following types:

$$\int \psi_a \psi_b d\tau \text{ and } \int \psi_a \boldsymbol{O} \psi_b d\tau$$

To develop an in-depth introduction to this topic see F. A. Cotton, *Chemical Applications of Group Theory*, 3rd Ed., J. Wiley and Sons, New York, 1990

where ψ_a and ψ_b are wavefunctions and $\boldsymbol{O}$ is an operator. The integrations are carried out over all space, $d\tau$ is a composite increment—$dxdydz$. These integrals are equal to zero unless they are invariant to all the symmetry operations of the point group. In group theoretical terms this corresponds to the integral belonging to the totally symmetric representation of the relevant point group. The symmetry properties of the wavefunctions mirror those of the irreducible representations and therefore the integral $\int \psi_a \psi_b d\tau$ corresponds in symmetry terms to the *direct product* of the characters of the irreducible representation corresponding to ψ_a and ψ_b, i.e. Γ_a and Γ_b.

Specifically, the characters of the representations of a direct product are equal to the products of the characters of the representations based on the individual sets of functions.

For example, for the C_{3v} point group the relevant part of the character table is:

C_{3v}	E	$2C_3$	$3\sigma_v$
A_1	1	1	1
A_2	1	1	−1
E	2	−1	0

Transformation properties of operators

Operator, $\mathcal{P}$	Transforms as
Electric dipole moment (vector)	x, y, z
Electric polarizability	x^2, y^2, z^2 xy, xz, yz
Magnetic dipole moment (axial vector)	R_x, R_y, R_z
Electric quadrupole moment	x^2, y^2, z^2 xy, xz, yz with the constraint that $x^2 + y^2 + z^2 = 0$

and the direct product table can be constructed from the following multiplications:

		E	$2C_3$	$3\sigma_v$		
$A_1 \times A_2$	≡	$1 \times 1 = 1$	$1 \times 1 = 1$	$1 \times -1 = -1$	≡	A_2
$A_2 \times E$	≡	$1 \times 2 = 2$	$1 \times -1 = -1$	$0 \times 1 = 0$	≡	E
$E \times E$	≡	$2 \times 2 = 4$	$-1 \times -1 = 1$	$0 \times 0 = 0$	≡	$A_1 + A_2 + E$

The irreducible representations which contribute to the reducible representations resulting from the direct product may be derived using equation (G.1). Alternatively a direct product table such as the one given below may be constructed.

Transformation properties of molecular translations and rotations

Molecular translations transform as	x, y, z
Molecular rotations transform as	R_x, R_y, R_z

C_{3v}	A_1	A_2	E
A_1	A_1	A_2	E
A_2	A_2	A_1	E
E	E	E	$A_1 + A_2 + E$

Such direct product tables lead to the following important generalization—the representation of a direct product contains the totally symmetric representation only if Γ_A and Γ_B belong to the same irreducible representation.

It follows that $\int \psi_A \psi_B d\tau \neq 0$ if Γ_A and Γ_B belong to the same irreducible representation.

The integral $\int\psi_A O\psi_B d\tau$ *is non-zero only if the representation of the direct product of any two functions is identical to the representation of the third component.*

Bonding implications of direct products

It is apparent that in a molecule the integrals $\int\psi_A\psi_B d\tau$ and $\int\psi_A H\psi_B d\tau$ are only zero if ψ_A and ψ_B belong to the same irreducible representation. The Hamiltonian operator, H, which is the energy operator belongs to the totally symmetric representation since if ψ is an eigenfunction of H, then $H\psi = E\psi$ and $\Gamma(H) \times \Gamma(\psi) = \Gamma(E) \times \Gamma(\psi)$ and $\Gamma(H)$ must be the same as $\Gamma(E)$ which is totally symmetric, i.e. the energy of the molecule must be invariant to symmetry operations. The generalization that only orbitals which belong to the same irreducible representation overlap and interact to form molecular orbitals delocalized over both centres follows directly from this analysis. The molecular orbital diagrams in Figs G.5 and G.6 were constructed on the basis of this initial assumption.

Direct products and selection rules for infrared and Raman spectroscopy

For a simple harmonic oscillator the permitted energy levels are given by:

$$E_i(n_i) = (n_i + \tfrac{1}{2})h\nu_i$$

where ν_i is the fundamental vibration frequency and n_i is the number of quanta in the ith oscillator. The wave function which describes the vibrational state of a molecule is given by:

$$\Psi_{vib} = \sum_{i=1}^{x} \psi_i(n_i) \quad (x = 3N-6 \text{ for a non-linear molecule or } 3N-5 \text{ for a linear molecule})$$

When $n_i = 0$ for each of the modes in their ground states the representation associated with Ψ_{vib} is totally symmetric. According to the Born-Oppenheimer approximation the electronic, vibrational, and rotational wave functions are separable and it is valid to consider the vibrational wave function in isolation. The probability p of a transition from one vibrational state to another is given by the square of the integral $\int\Psi_2\mathcal{P}\Psi_1 d\tau$, where $\mathcal{P}$ is the operator responsible for the transition between Ψ_1 (ground state) and Ψ_2 (excited state).

Whether the integral is zero or non-zero can be determined from symmetry theory by deriving the character of the triple product of the individual characters of the components of the integral. The triple product can be expressed as:

$$\Gamma(\text{integral}) = \Gamma(\Psi_1) \times \Gamma(\mathcal{P}) \times \Gamma(\Psi_2)$$

An infrared-induced transition is governed by the electric vector operator which has the same transformation properties as x, y, and z. Therefore:

$$\Gamma(\mathcal{P}, \text{i.r.}) = \Gamma(x, y, z)$$

A Raman-induced transition depends on the change in polarizability of a molecule under the effect of visible light and

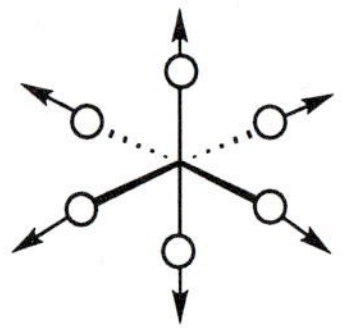

Symmetric stretch: a_{1g} Raman

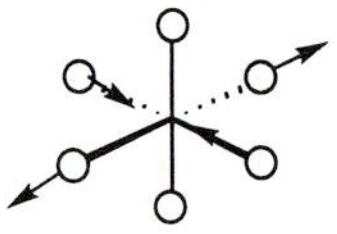

e_g Raman 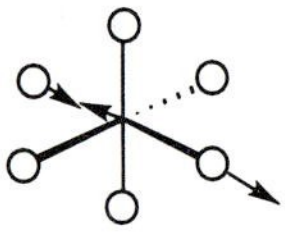 t_{1u} i.r.

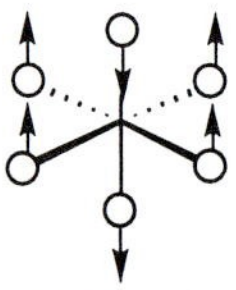

t_{1u} i.r. 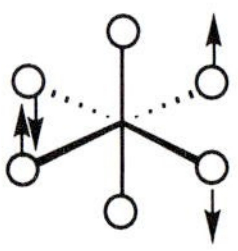 t_{2u} inactive
Out-of-plane bending

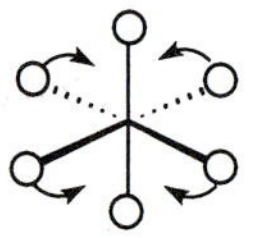

t_{2g} Raman, in-plane bending

Illustration of the normal vibrational modes for an octahedral ML_6 complex. The i.r. and Raman activities of the modes are also indicated. Note that since the molecule has a centre of symmetry the Raman and i.r. modes are mutually exclusive

$$\Gamma(\mathcal{P}, \text{Raman}) = \Gamma(x^2, y^2, z^2, xy, xz, yz)$$

It follows that a specific vibrational mode of a molecule is only infrared active (i.e there is a non-zero value of the integral for transitions from the ground state) if it belongs to the same irreducible representation as x, y, and z and Raman active if it belongs to the irreducible representations associated with x^2, y^2, z^2, xy, xz, and yz.

If a molecule has a centre of symmetry all the representations associated with x, y, and z are ungerade (u) and those associated with x^2, y^2, z^2, xy, xz, and yz are gerade (g). Therefore, only ungerade vibrations of the molecule can possibly be infrared active and gerade vibrations Raman active. Therefore, no absorptions corresponding to fundamental vibrations may be both infrared and Raman active. This is the basis of the *mutual exclusion rule*.

On page 45 it was noted that the CO stretching modes in the centrosymmetric $Mo(CO)_6$ molecule transform as $a_{1g} + e_g + t_{1u}$. It follows that the a_{1g} and e_g modes are exclusively Raman active and the t_{1u} mode is exclusively infrared active.

For a tetrahedral EH_4 molecule the E–H bond stretching modes transform as $a_1 + t_2$. The a_1 mode is exclusively Raman active since $x^2 + y^2 + z^2$ transforms as a_1 (see page 45). The t_2 mode is both infrared and Raman active since x , y , z and xz, yz, xy transform as t_2.

Direct products and selection rules for electronic transitions

The electric vector determines whether the absorption of electromagnetic radiation is symmetry allowed or forbidden. The transition probability is given by the square of the integral:

$$\int \Psi_{ex} \mathcal{P} \Psi_g \, d\tau$$

where Ψ_g and Ψ_{ex} are the wave functions of the ground and excited states respectively and $\mathcal{P}$ is the electric vector operator responsible for the transition. In symmetry terms the relevant direct product is:

$$\Gamma(\Psi_g) \times \Gamma(\mathcal{P}) \times \Gamma(\Psi_{ex})$$

For an allowed transition the triple product must be or contain the totally symmetric representation. $\Gamma(\mathcal{P})$ must transform as x, y, and z, since it is the electric vector operator.

The irreducible representations associated with Ψ_g and Ψ_{ex} may be derived from the representations of the specific orbitals occupied. The representation of an electron in an orbital corresponds to the representation of that orbital and the representation of the wave function of several electrons is the direct product of the representations for the individual electrons.

If the molecule has a ground state where the electrons are all paired and all the orbitals are fully occupied then the Ψ_g corresponds to the totally symmetric representation. Therefore, complete filled orbitals can be ignored when establishing the symmetry of a state. For example, the ground state configuration $(\psi_1)^2(\psi_2)^2(\psi_3)^1$ corresponds to the product:

$$\Gamma(\psi_1) \times \Gamma(\psi_1) \times \Gamma(\psi_2) \times \Gamma(\psi_2) \times \Gamma(\psi_3)$$

However, since ψ_1 and ψ_2 are fully occupied the product can be abbreviated to $\Gamma(\psi_3)$.

A hole in an orbital set has the symmetry transformation properties as a single electron occupying that set of orbitals would. Therefore, for example, in symmetry terms t_2^5 may be replaced by t_2^1 and e^3 by e^1.

The tetrahedral EH_4 molecule with 8 valence electrons, for example, has a ground state corresponding to $(a_1)^2(t_2)^6$ which belongs to the totally symmetric representation. According to the qualitative molecular orbital diagram, the lowest lying excited states will be associated with the following electron excitations:

$\Psi_{ex}(1) = (a_1)^2(t_2)^5(a_1^*)^1$ and $\Psi_{ex}(2) = (a_1)^2(t_2)^5(t_2^*)^1$ and their symmetry transformation properties correspond to:

$$\Gamma(\Psi_{ex}(1)) = \Gamma(t_2) \times \Gamma(a_1) = \Gamma(t_2) \text{ and}$$
$$\Gamma(\Psi_{ex}(2)) = \Gamma(t_2) \times \Gamma(t_2) = \Gamma(t_2) + \Gamma(t_1) + \Gamma(e) + \Gamma(a_1)$$

The transition probability for the first excitation is $\int \Psi_{ex}(1)\mathcal{P}\Psi_g d\tau$ and has the symmetry property: $\Gamma(\Psi_{ex}(1)) \times \Gamma(\mathcal{P}) \times \Gamma(\Psi_g) = t_2 \times t_2 \times a_1$, since $\mathcal{P}$ transforms as x, y, and z (t_2). The product has a totally symmetric component because $t_2 \times t_2 = a_1 + e + t_1 + t_2$ and $a_1 \times a_1 = a_1$ and is therefore an allowed transition.

The second transition also has a t_2 component and is also allowed. Consequently both transitions are symmetry allowed.

The spin selection rule dictates that the transition must retain the same spin multiplicity and therefore the symmetry allowed transitions described above may be defined as follows:

$$^1A_1 \rightarrow {}^1T_2 \ (t_2 \rightarrow a_1^*)$$
$$^1A_1 \rightarrow {}^1T_2 \ (t_2 \rightarrow t_2^*)$$

It is conventional to describe electronic states by upper case Schönflies symbols and to use lower case symbols for orbitals.

Some transitions, although symmetry forbidden, are observed albeit with low intensities because of vibronic coupling effects. The Born–Oppenheimer approximation separates the electronic and vibrational parts of the wave function, but if there is some coupling between them the more appropriate direct product for the transition moment is:

$$\Gamma(\Psi_g(\text{elec})) \times \Gamma(\Psi_g(\text{vib})) \times \Gamma(\mathcal{P}) \times \Gamma(\Psi_{ex}(\text{elec})) \times \Gamma(\Psi_{ex}(\text{vib}))$$

where (elec) and (vib) represent electronic and vibrational wave functions. If the molecule is simultaneously in its electronic and vibrational ground state vibronic coupling will only lead to a transition probability if $\Gamma(\mathcal{P}) \times \Gamma(\Psi_{ex}(\text{elec})) \times \Gamma(\Psi_{ex}(\text{vib}))$ is totally symmetric.

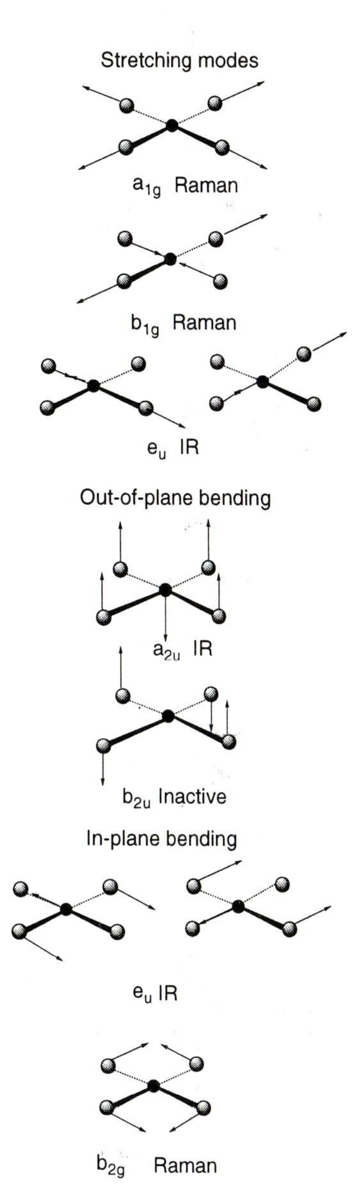

Illustration of the normal vibrational modes for a square planar ML_4 complex. The i.r. and Raman activities of the modes are also indicated. Note that since the molecule has a centre of symmetry the Raman and i.r. modes are mutually exclusive

Stretching modes

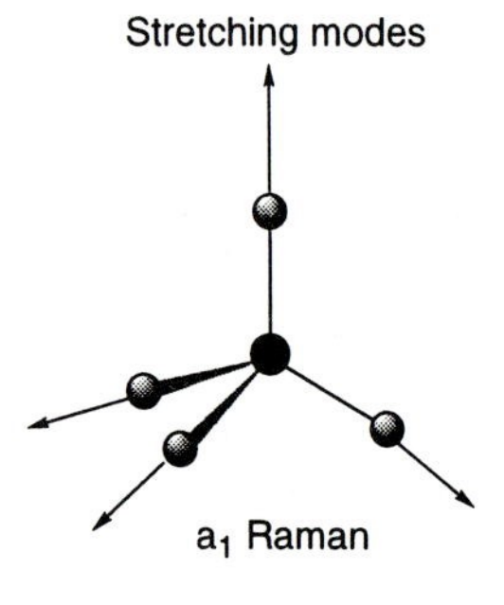

a_1 Raman

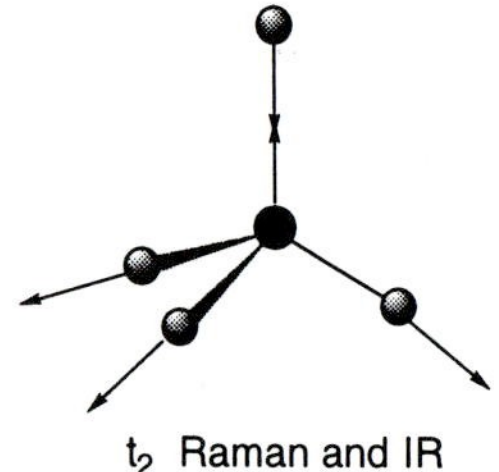

t_2 Raman and IR

Bending modes

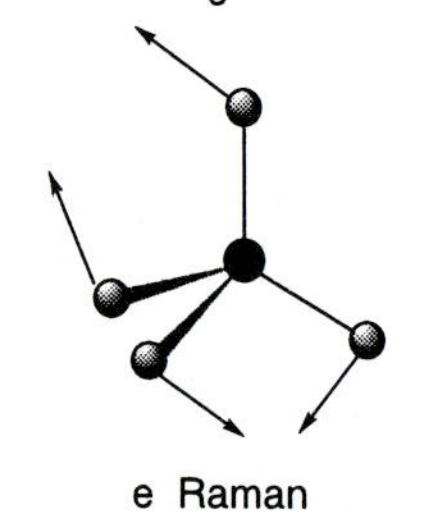

e Raman

t_2 Raman and IR

Illustration of the normal vibrational modes for a tetrahedral ML_4 complex. The i.r. and Raman activities of the modes are also indicated. Since the tetrahedron has no centre of symmetry the t_2 modes are Raman and i.r. active

For example, for an octahedral transition metal complex with a d^6 low-spin configuration the relevant ground state electronic populations are derived from Fig. G.6 are:

$$(a_{1g})^2(t_{1u})^6(e_g)^4(t_{2g})^6$$

The molecular wave function is therefore totally symmetric and spin paired, i.e. ${}^1A_{1g}$. The lowest energy excitations to the $e_g{}^*$ molecular orbitals lead to the following states:

$${}^1T_{1g},\ {}^1T_{2g},\ {}^3T_{1g},\ {}^3T_{2g}\ ((t_{2g})^5 \times (e_g{}^*)^1) \equiv (t_{2g} \times e_g) = t_{1g} + t_{2g}$$

since a hole in an orbital set has the same symmetry transformation properties as a single electron in that set. Therefore, the spin-allowed transitions for the octahedral complex are:

$${}^1A_{1g} \rightarrow {}^1T_{1g}$$
$${}^1A_{1g} \rightarrow {}^1T_{2g}$$

These transitions are forbidden based on the electric vector operator since

$$\Gamma(\Psi_g) \times \Gamma(\boldsymbol{\mathcal{P}}) \times \Gamma(\Psi_{ex}) = A_{1g} \times T_{1u} \times T_{1g}\ (\text{or } T_{2g})$$

has no totally symmetric component.

If the transition is vibronically coupled to a mode of t_{1u} symmetry the direct product:

$$\Gamma(\Psi_g) \times \Gamma(\boldsymbol{\mathcal{P}}) \times \Gamma(\Psi_{ex}(\text{elec})) \times \Gamma(\Psi_{ex}(\text{vib}))$$
$$= A_{1g} \times T_{1u} \times T_{1g}\ (\text{or } T_{2g}) \times T_{1u}$$

does have a totally symmetric component since

$$T_{1u} \times T_{1u} = A_{1g} + T_{1g} + T_{1g} + T_{2g}$$
$$\text{and } A_{1g} \times T_{2g} \times T_{2g} \text{ and } A_{1g} \times T_{1g} \times T_{1g}$$

both having totally symmetric components in their direct products.

Therefore, such transitions are observed because the coupling of the vibrational modes with the electronic wavefunctions is sufficiently effective to remove the centre of symmetry within the molecule at the instant when the electronic transition occurs. The extinction coefficients of the transitions, which are related to the transition probabilities, remain smaller than those for electronically allowed transitions.

π-orbitals in coordination and polyhedral molecules

On pages 43 and 44 the derivation of the symmetries of the σ-type orbitals of coordination complexes were obtained using the relevant character tables. Direct products may be used to establish the symmetries of the π-type linear combinations of ligand orbitals. Specifically the following direct product multiplications define the symmetries of the ligand π-orbitals:

$$\Gamma_\sigma + \Gamma_\pi = \Gamma_\sigma \times \Gamma_{x,y,z}$$

$$\text{and } \Gamma_\pi = \Gamma_\sigma \times \Gamma_{x,y,z} - \Gamma_\sigma$$

where $\Gamma_{x,y,z}$ correspond to the irreducible representation for the vectors x, y, and z and Γ_σ and Γ_π correspond to the irreducible representations for the σ and π linear combinations of atomic orbitals.

For an octahedral complex:

$$\Gamma_\sigma = a_{1g} + e_g + t_{1u} \text{ and } \Gamma_{x,y,z} = t_{1u}$$

$$\begin{aligned}\Gamma_\sigma + \Gamma_\pi &= (a_{1g} + e_g + t_{1u}) \times t_{1u}\\ &= t_{1u} + (t_{1u} + t_{2u}) + (t_{1g} + t_{2g} + a_{1g} + e_g)\end{aligned}$$

$$\Gamma_\pi = t_{1u} + t_{2u} + t_{1g} + t_{2g}$$

Representative examples of the linear combinations of atomic orbitals which have these symmetry transformation properties are illustrated in Fig. G.9.

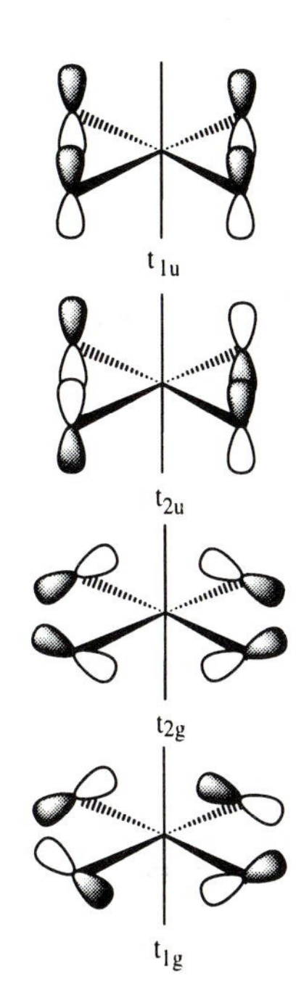

Fig. G.9 Examples of linear combinations of π-type orbitals in an octahedral complex. The tangential orbitals in an octahedral polyhedral molecule have the same symmetry transformation properties

The application of the direct product to a tetrahedral ML_4 complex, where $\Gamma_\sigma = a_1 + t_2$ leads to the following representations for Γ_π:

$$\Gamma_\sigma + \Gamma_\pi = (a_1 + t_2) \times t_2 = t_2 + a_1 + t_2 + t_1 + e$$
$$\Gamma_\pi = t_2 + t_1 + e$$

Stretching and bending modes

The direct products may also be used to make a full vibrational analysis of the stretching and bending vibrational modes in a molecule. The stretching modes have the same symmetry transformation properties as Γ_σ and therefore the 3N displacement vectors (shown in Fig. G.10 for an octahedral molecule) for the possible translations of the atoms on the periphery of the molecule are given by: $\Gamma_\sigma + \Gamma_\pi = \Gamma_\sigma \times \Gamma_{x,y,z}$. The displacement vectors for the central atom transform as x, y, and z therefore for a molecule the transformation properties of all the displacement vectors are: $\Gamma_\sigma \times \Gamma_{x,y,z} + \Gamma_{x,y,z}$

For example, for the octahedron:

$$\Gamma_\sigma \times \Gamma_{x,y,z} + \Gamma_{x,y,z} = t_{1u} + (t_{1u} + t_{2u}) + (t_{1g} + t_{2g} + a_{1g} + e_g) + t_{1u}$$

The (3N – 6) vibrational modes may be obtained from these by subtracting the relevant translational and rotational modes of the whole molecule. The translational modes transform as $\Gamma_{x,y,z}$ in the character table and the rotational modes Γ_{R_x,R_y,R_z}. Specifically for the octahedron these are t_{1u} and t_{1g} respectively, therefore the genuine vibrational modes transform as:

$$t_{1u} + t_{2u} + t_{2g} + a_{1g} + e_g + t_{1u}$$

The stretching modes are: a_{1g} (Raman) + e_g (Raman) + t_{1u} (Infrared)

The bending modes are: t_{1u} (Infrared), t_{2u} neither infrared nor Raman active), and t_{2g}(Raman active).

Diagrams of the vibrational modes for octahedral (p. 48), square planar (p. 49), and tetrahedral (p. 50) molecules are illustrated in the margins of the pages noted above.

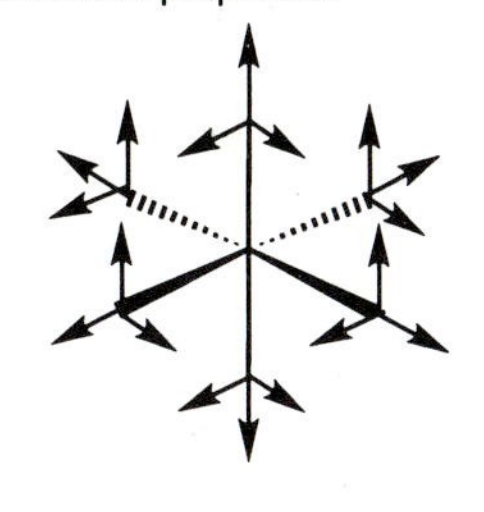

Fig. G.10 The displacement vectors for the ligand donor atoms (top) and the central metal ion (bottom) for an octahedral complex

I

Isolobal analogy

Isolobal analogy

The isolobal analogy provides a very useful methodology for rationalizing the structures of polyhedral molecules having both main group and transition metal fragments. Fig. I.1 illustrates a series of tetrahedral molecules where C–H fragments are progressively being replaced by $Co(CO)_3$ fragments. Therefore, the two fragments must have similar bonding capabilities. Specifically, both fragments have three out-pointing hybrid orbitals, for C–H based on sp^3 hybrids, and for $Co(CO)_3$ based on d^2sp^3 hybrids each of which is occupied by a single electron. Fragments which exhibit similar bonding capabilities are described as *isolobal* and the symbol ◄─o─► is used to indicate this similarity. Further examples of such isolobal relationships are illustrated in Fig. I.2.

It is significant that with these metal carbonyl fragments of the later transition metals the d_δ orbitals are not used and therefore they form a maximum of three out-pointing hybrids, thus making them analogous to the main group fragments. The fragments illustrated in Fig. I.2 include examples where they provide two or one orbitals for bonding.

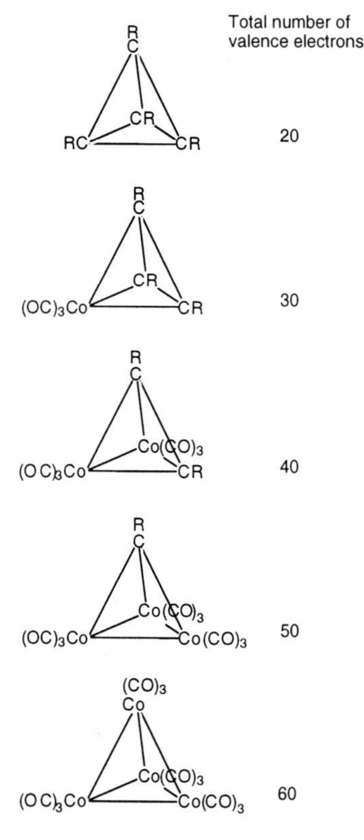

Fig. I.1 An isostructural series of molecules with main group and transition metal fragments at the vertices of a tetrahedron

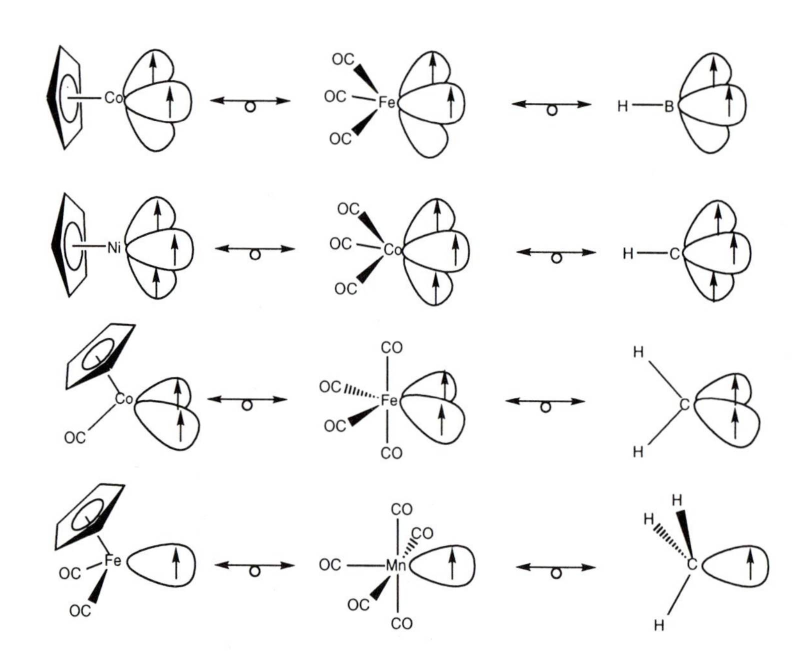

Fig. I.2 Examples of the isolobal analogy

It is noteworthy that the total number of valence electrons in these isolobal fragments differ by 10. Therefore, a series of isostructural polyhedral molecules such as that shown in Fig. I.1 show a similar increment as the number of metal atoms is increased.

More examples of the application of the isolobal analogy to polyhedral molecules are given in Fig. I.3.

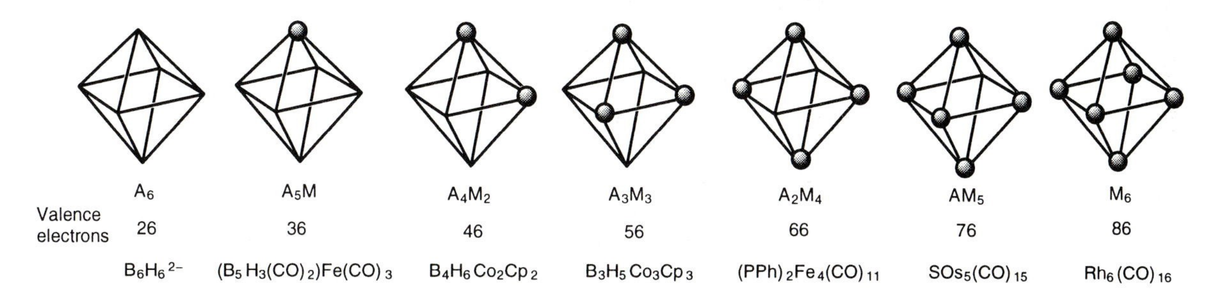

Fig. I.3 Examples of series of *closo*-octahedral cluster compounds starting with $B_6H_6^{2-}$ and ending with $Rh_6(CO)_{16}$. The increment of ten valence electrons each time a metal atom is introduced into the cage is particularly noteworthy

Since the *isolobal* analogy is clearly linked to the Effective Atomic Number Rule it has to be modified and indeed in some cases abandoned for classes of complexes where the rule is inapplicable. For example, the platinum and coinage metals which show a preference for forming 16 and 14 electron complexes in their lower oxidation states have isolobal relationships which reflect these preferences. A T-shaped d^8 ML_3 complex is thereby *isolobal* with CH_3^+ and an angular d^8 ML_2 fragment is *isolobal* with CH_2^{2+}. Similarly a d^{10} ML fragment, e.g. $Au(PPh_3)^+$ is isolobal with CH_3^+ and H^+. These relationships are illustrated in Fig. I.4. The *isolobal* analogy is not applicable to metal fragments of the earlier transition metals with π-donor ligands, e.g. $ReCl_4^-$ where the metal uses d_δ orbitals in bonding.

For a comprehensive account of the *isolobal* analogy and fragment molecular orbital analysis see T. A. Albright, J. K. Burdett, and M-H. Whangbo, *Orbital interactions in chemistry*, J. Wiley and Sons, New York, 1985

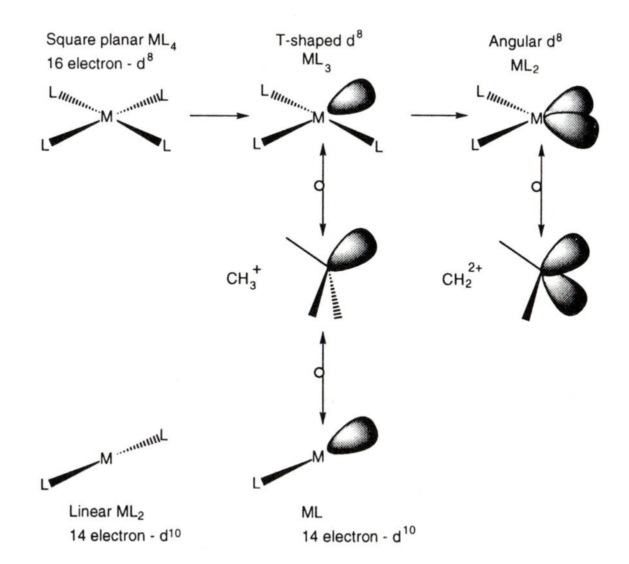

Fig. I.4. Isolobal relationships for 16 and 14 electron metal complexes

L

Ligand substitution reactions

Ligand substitution reactions

Following the nomenclature proposed by Langford and Gray substitution reactions of inorganic reactions are classified as *associative* (*A*), *interchange* (*I*) and *dissociative* (*D*). In an *associative* mechanism, A, the entering ligand, Y, binds to the central atom, M, before any significant bond weakening to the leaving group, X, occurs (Fig. L.1(a)). The formation of the new M–Y bond leads to an *intermediate* with a higher coordination number and the reaction profile involves two transition states—the first leading to the formation of the intermediate and the second the loss of X to form the substituted product MX. For a *dissociatve mechanism*, *D*, the ligand, X, departs before the entering group, Y, forms a significant bond with M and an *intermediate* with a lower coordination number is formed. The formation of such an intermediate also results in two peaks in the reaction profile (Fig. L.1(c)), but the activation energy leading to the intermediate is larger than for its decomposition. The depth of the depression between the two peaks determines the steady state concentration of the intermediate and if it is significant the intermediate may be detected by spectroscopic means. Experimentally the detection of an intermediate is important for identifying a mechanism as *A* or *D*.

If the bond making and breaking processes occur simultaneously then the mechanism of the resulting concerted reaction is described as *interchange, I*. The reaction profile (Fig. L.1(b)) involves only a single peak and no intermediate is formed.

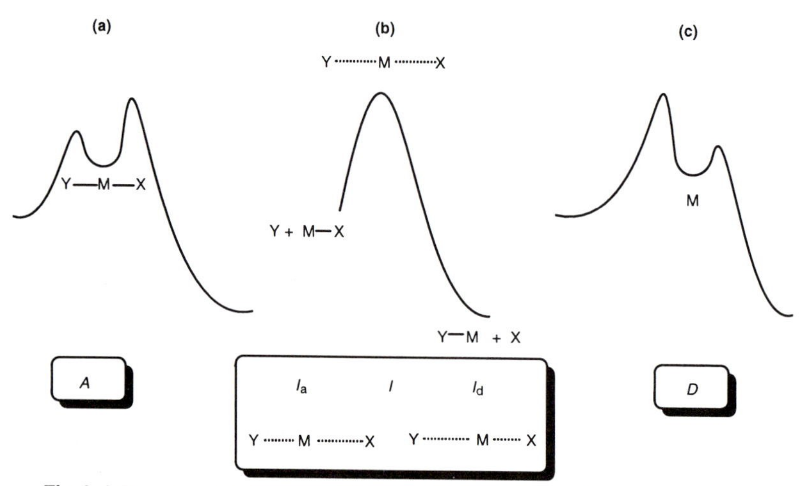

Fig. L.1 Nomenclature for ligand substitution reactions: Y + M—X → Y—M + X

In practice the study of inorganic reactions has led to a whole spectrum of behaviour from *A* to *I* to *D*, and suggested the need to subdivide the *I* category: If the bond making and breaking process are not exactly synchronous an intermediate may arise but its stabilization energy may not be sufficiently large for it to form in sufficient concentrations to be easily detected spectroscopically. If the bond making occurs somewhat more quickly than the bond breaking the mechanism is described as I_a, whereas if the bond breaking occurs more quickly than the bond making the mechanism is described as I_d.

Besides the detection of an intermediate the volume of activation for the reaction, which is obtained by studying the effect of pressure on the reaction rate, may provide an indication of whether the mechanism is best classified as A, I_a, I, I_d or D. Dissociative reactions have large positive volumes of activation because the formation of the intermediate requires the release of a ligand, whereas associative reactions have negative volumes of activation. For interchange reactions the volumes of activation are close to zero, with I_d slightly positive and I_a slightly negative. The entropies of activation can also provide a means of distinguishing between the mechanisms (see Table L.1 for examples.

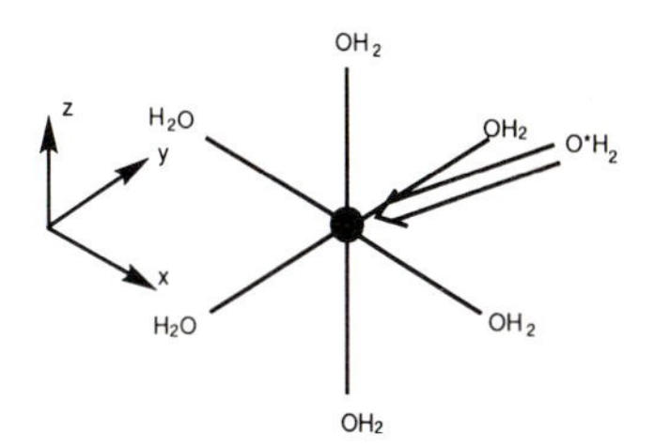

Fig. L.2 The nucleophilic substitution by the labelled water molecule is favoured along the three-fold axis of the octahedron for steric reasons. The metal d_{z^2} and $d_{x^2-y^2}$ orbitals point directly towards the ligands and the d_{xy}, d_{xz}, and d_{yz} orbitals have their maxima in between the axes. If one or more of these orbitals are empty the lone pair on the incoming water molecule can enter into a favourable two-orbital two-electron stabilizing interaction. This encourages an I_a mechanism. However, if they are fully occupied then there is a repulsive interaction between the lone pair and the filled d orbital and an I_a mechanism is less favoured

Specific examples

The exchange reaction:

$$[M(OH_2)_6] + O^*H_2 \rightarrow [M(OH_2)_5(O^*H_2)] + OH_2$$

$$I_a\ M = Ti^{II}, V^{III}, Cr^{III}, V^{II};\ I/I_a\ Mn^{II}, Fe^{II}, \text{and } Fe^{III};\ I_d\ M = Co^{II} \text{ and } Ni^{II}$$

occurs via an interchange mechanism, but the volume of activation varies systematically across the first transition series from $-12\ cm^3\ mol^{-1}$ for metal ions on the left hand side of the transition series (suggesting an I_a mechanism) to $+7\ cm^3\ mol^{-1}$ for metal ions on the right hand side of the transition series (suggesting an I_d mechanism). See Fig. L.2 for an interpretation of this change in mechanism.

Associative reactions

Substitution reactions of tetrahedral compounds of the Group 14 and 15 elements, e.g. MR_3Cl (M = Si, Ge, and Sn) and phosphine substitution reactions of tetrahedral transition metal complexes, e.g. $MCl_2(PR_3)_2$ (M = Co or Ni) proceed via an associative mechanism.

Substitution reactions of square planar d^8 complexes. These may involve either direct attack by the ligand or a nucleophilic solvent molecule. The resulting five coordinate intermediate is sometimes detected (A mechanism), but more usually not (I_a) mechanism. The volumes of activation lie in the range -10 to $-5\ cm^3\ mol^{-1}$.

18 electron organo-transition metal complexes which have ligands capable of functioning ether as n or $(n - 2)$ electron donors, e.g. NO and C_5H_5, undergo associative substitution reactions.

Table L.1 Some entropies ($J\ K^{-1}\ mol^{-1}$) and volumes ($cm^3\ mol^{-1}$) of activation for the exchange of dimethylformamide (dmf) in octahedral high-spin complexes

	$\Delta S^\ddagger$	$\Delta V^\ddagger$
$[Al(dmf)_6]^{3+}$	28	+13.7
$[Cr(dmf)_6]^{3+}$	−43	−6.3
$[Fe(dmf)_6]^{3+}$	−69	−0.9
$[Fe(dmf)_6]^{2+}$	+14	+8.5
$[Ni(dmf)_6]^{2+}$	+34	+9.1

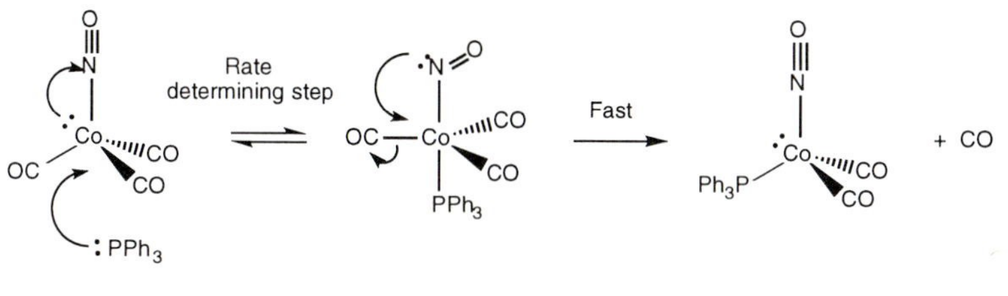

Rate determining step

Fast

+ CO

Rate determining step
Fast
OC Co CO
: PPh_3
Ph_3P Co CO CO
η^3
Ph_3P Co CO
η^5
+ CO

The starting complex, the final product, and the transition state complex thereby conform to the EAN rule.

The angular overlap model may be used to calculate the relative energies of octahedral and capped octahedral geometries. The energy difference is smallest for d^2 and low spin d^4complexes

Dissociative reactions

Substitution reactions of 18 electron organo-transition metal complexes with unambiguous electron donating characteristics, e.g. CO, PR_3, NH_3, generally proceed via a dissociative mechanism. For example, the reactions of $Ni(CO)_4$, $Ni(P(OEt)_3)_4$, $Cr(CO)_6$, and $[Rh(NH_3)_6]^{3+}$ generally have a first order rate law (*rate = k[complex]*). If a lower coordination number intermediate is detected and there is a large positive volume of activation the mechanism is *D*. The absence of an intermediate and with a rate law of the type: rate = *k*[complex][ligand], together with a small positive volume of activation, suggests I_d.

Table L.2 Rate constants, *k*, for exchange of OH_2 at 25°C

Metal ion	d^n	k/s^{-1}
$[Na(OH_2)_6]^+$	0	8×10^9
$[Mg(OH_2)_6]^{2+}$	0	1×10^5
$[Ca(OH_2)_6]^{2+}$	0	2×10^8
$[Sr(OH_2)_6]^{2+}$	0	4×10^8
$[Al(OH_2)_6]^{3+}$	0	1.8
$[Ga(OH_2)_6]^{3+}$	0	1×10^3
$[In(OH_2)_6]^{3+}$	0	2×10^5
$[Cr(OH_2)_6]^{3+}$	3	3×10^{-6}
$[Cr(OH_2)_6]^{2+}$	4	7×10^9
$[Mn(OH_2)_6]^{2+}$	5	3×10^7
$[Fe(OH_2)_6]^{3+}$	5	3×10^3
$[Fe(OH_2)_6]^{2+}$	6	3×10^6
$[Co(OH_2)_6]^{2+}$	7	1×10^6
$[Ni(OH_2)_6]^{2+}$	8	3×10^4
$[Cu(OH_2)_6]^{2+}$	9	8×10^9
$[Zn(OH_2)_6]^{2+}$	10	2×10^7

Rates of ligand substitution of metal complexes

Table L.2 summarizes the rate constants for water exchange in a range of complexes. Clearly the $M^{2+}(aq)$ ions are much more labile than the $M^{3+}(aq)$ ions and the rates increase as the size of the ion increases for the non-transition elements. Within the $M^{2+}(aq)$ series the rates depend very much on the d configuration of the ions. The most inert ion is $Ni^{2+}(aq)$, d^8, which has the maximum ligand field stabilization energies for a high spin complex. Either ligand addition or loss would result in some loss of this stabilization energy. d^3 and low spin d^6 complexes are also inert because of the loss of ligand stabilization energy when achieving the transition state.

The d^5, d^6, and d^7 ions have exchange rates of approximately 10^6–10^7 s^{-1} which is in the range anticipated for M^{2+} aqueous ions on the basis of size effects. For these ions the ligand field stabilization energies for the octahedron are zero (Mn^{2+}, d^5) or small (Fe^{2+}, d^6, Co^{2+}, d^7). In contrast for the d^4 and d^9 complexes of Cr^{2+} and Cu^{2+} are particularly labile (rates ≈ 10^9 s^{-1}) because complexes with these electronic configurations are known to have non-regular octahedral geometries. Generally, these complexes have tetragonally distorted geometries. This distortion arises from a stabilization of the d_{z^2} orbital in the e_g set of orbitals of an octahedron.

For a fuller discussion of ligand field stabilization energies see page 9 and R. A. Henderson, *The Mechanisms of Reactions at Transition Metal Sites*, OUP, Oxford, 1993, p 27–28.

Magnetic phenomena (co-operative)

The origins of diamagnetism and paramagnetism were discussed in *Essentials of Inorganic Chemistry* 1 and such magnetic properties of an individual complex can be related to those observed for a solid sample containing many billions of molecules. However, there are other compounds where the electron spins on any particular molecule are influenced by those on neighbouring molecules and the observed magnetic behaviour is therefore a collective phenomenon.

A conventional paramagnetic substance has a molar susceptibility, χ, which is inversely proportional to temperature (the Curie law states $\chi = C/T$ where C is the Curie constant) leading to the plot shown in Fig. M.1(a). In a *ferromagnetic* substance the magnetic spins on one molecule prefer to line up parallel to those on neighbouring molecules. This preference may be sustained over many thousands of unit cells to create a *magnetic domain.* A ferromagnetic sample has the characteristic χ vs. T plot shown in Fig. M.1(b), where below the Curie Temperature T_c the susceptibility of the sample is increased relative to that expected on the basis of the Curie Law, because of the alignment of the spins.

In an *anti-ferromagnetic* substance the magnetic spins on one molecule prefer to lie anti-parallel to those of neighbouring molecules and consequently the susceptibility below the Néel temperature falls relative to that expected from the Curie Law (Fig. M.1(c)).

In *ferrimagnetic* substances a net magnetic ordering is observed below the Néel Temperature, because the spins align in an anti-parallel fashion, but since there is more than one type of magnetic centre in the substance the spins do not cancel out completely. The sample therefore has a net magnetic susceptibility even at absolute zero.

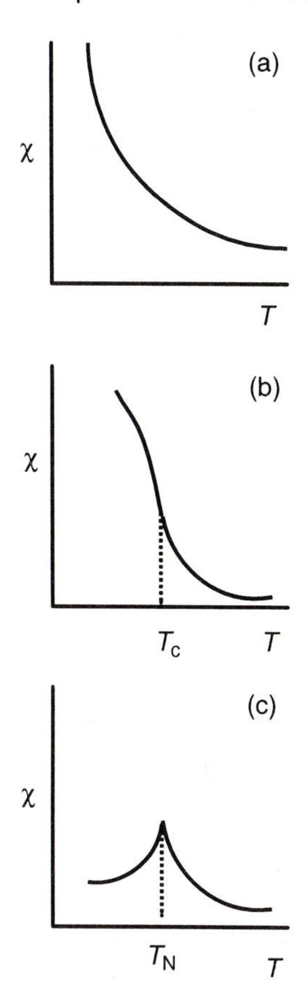

Fig. M.1 Characteristic molar susceptibility – temperature plots for (a) paramagnetic, (b) ferromagnetic and (c) antiferromagnetic substances. In (b) ferromagnetic behaviour occurs below the Curie temperature, T_C, and in (c) anti-ferromagnetic behaviour occurs below the Néel temperature, T_N

Transition metal oxides, MO, all have the sodium chloride structure although at times they are not simple stoichiometric compounds. In the sodium chloride structure the octahedral environment around the metal leads to the e_g–t_{2g} splitting discussed previously (p. 6). The non-bonding t_{2g} set point towards the matching orbitals on the adjacent metal ions and the overlap between the t_{2g} orbitals is small but sufficiently significant for a band structure to develop.

If the ordering occurs in such a way that the electron spins on the metals all orientate preferentially in a parallel fashion *ferromagnetism* results. If the electron spins on the metals preferentially align in an anti-parallel fashion, as shown in Fig. M.2, *anti-ferromagnetism* results. If the interactions which result in these alignments are strong the transition from ferromagnetic or anti-ferromagnetic behaviour to paramagnetic behaviour occurs at a high temperature, whereas if they are weak very low temperatures are required for this transition.

An important mechanism by which the spins of metal ions align in solids is the *super exchange mechanism.* In the metal oxides MO which have the sodium chloride structures the electrons in the e_g orbitals can interact with the orbital pair in the p orbitals of the oxide to polarize the electron spin on the adjacent metal ion in an anti-parallel fashion, i.e. leading to an anti-ferromagnetic interaction. Consequently the metal oxides MnO, Fe_{1-x}, CoO,

For a more detailed discussion see P. A. Cox, *Transition Metal Oxides*, Clarendon Press, Oxford, 1992

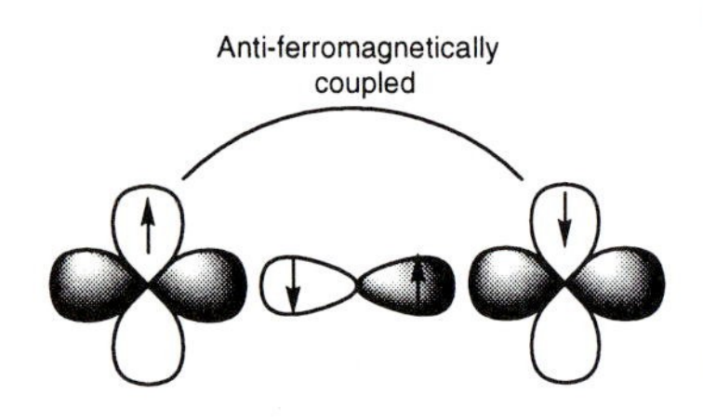

Fig. M.2 A diagram illustrating the anti-ferromagnetic coupling of metal orbitals via the spin polarization of the electrons in the oxygen 2p orbital

and NiO which have electrons occupying the e_g orbitals exhibit anti-ferromagnetic interactions. The corresponding Néel temperatures are:

MnO	$Fe_{1-x}O$	CoO	NiO
122 K	198 K	293 K	523 K

The increasing Néel temperature across the series reflects the stronger metal–ligand interactions across the series as the metal ions contract in size.

Multiple metal–metal bonds

The transition metals form a wide range of compounds which contain more than one metal atom. The particularly short metal-metal distances in some of these compounds suggest the presence of direct metal-to-metal bonding. The formation of multiple metal–metal bonds requires the effective overlap between the d orbitals of the metal atoms. As the metal oxidation state is increased the orbitals contract and their overlaps with the d orbitals of adjacent atoms diminish. The nodal properties of the d orbitals, as shown in Fig. M.3, allow overlaps between the d orbitals of two metal atoms which could result in a maximum metal–metal bond order of five, i.e. σ, two π, and two δ bonds.

Fig. M.3 Nodal properties of d orbitals relating to the *z* axis

In dimeric compounds the maximum bond order is not observed because one or more of the d orbitals are used in forming metal-to-ligand bonds and are unavailable for multiple metal-metal bond formation. For example, in $[Re_2Cl_8]^{2-}$, where the metals have a formal oxidation state of +3 (i.e. the metal ion configuration is d^4) the metal $d_{x^2-y^2}$ orbital is used in forming dsp^2 hybrids which are able to accept lone pairs of electrons from the four ligands in forming square planar $ReCl_4^-$ groups. The remaining four metal d orbitals overlap to form a quadruple metal–metal bond. The bonding and antibonding molecular orbitals associated with such a quadruple bond are illustrated in Fig. M.4. The δ component requires that the $ReCl_4^-$ fragments adopt an eclipsed conformation to achieve maximum overlap between the d_{xy} orbitals.

There are series of related dimeric molecules in which the formal metal–metal bond order changes with the population of the antibonding orbitals shown in Fig. M.4. Examples of such compounds with 8–16 electrons are given in Table M.1. The metal–metal bond length progressively increases as the formal metal–metal bond order decreases.

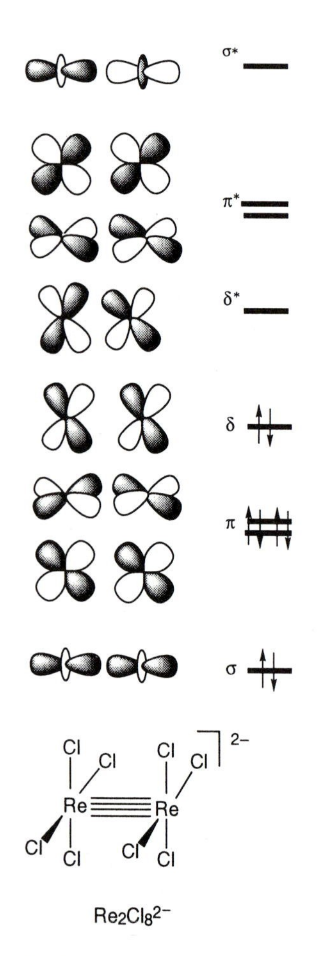

Fig. M.4 The molecular orbitals and structure of the $[Re_2Cl_8]^{2-}$ ion

For the early transition metals the metal–metal bond order is determined primarily by the number of d electrons associated with the metal ion in its formal oxidation state. A d^4 metal ion, e.g. Mo^{2+}, Tc^{3+}, or Re^{3+} is capable of forming a total of four metal–metal bonds.

Table M.1 Examples of dimeric metal–metal bonded compounds

Compound and the number of d electrons involved in metal–metal bonding		Formal bond order	M–M distance /pm	Electronic configuration
$[Re_2Cl_8]^{2-}$	8	4	224.1(7)	$\sigma^2\pi^4\delta^2$
$[Re_2Br_8]^{2-}$	8	4	222.8(4)	$\sigma^2\pi^4\delta^2$
$[Tc_2Cl_8]^{2-}$	8	4	214.7(4)	$\sigma^2\pi^4\delta^2$
$Re_2Cl_4(PMe_3)_4$	10	3	224.7(1)	$\sigma^2\pi^4\delta^2\delta^{*2}$
$Tc_2Cl_4(PMe_2Ph)_4$	10	3	212.8(1)	$\sigma^2\pi^4\delta^2\delta^{*2}$
$[Os_2Cl_8]^{2-}$	10	3	219.5(2)	$\sigma^2\pi^4\delta^2\delta^{*2}$
$[Ru_2(\mu_2\text{-acetato})_4(OH_2)_2]^+$	11	2.5	224.8(1)	$\sigma^2\pi^4\delta^2\delta^{*2}\pi^{*1}$
$[Ru_2(\mu_2\text{-acetato})_4(thf)_2]$	12	2	226.1(3)	$\sigma^2\pi^4\delta^2\delta^{*2}\pi^{*2}$
$[Rh_2(\mu_2\text{-acetato})_4(S\text{-thf})_2]$	14	1	241.3(1)	$\sigma^2\pi^4\delta^2\delta^{*2}\pi^{*4}$
$[Pd_2(\mu_2\text{-acetato})_4(OH_2)_2]^{2+}$	16	0	257.8(2)	$\sigma^2\pi^4\delta^2\delta^{*2}\pi^{*4}\sigma^{*2}$

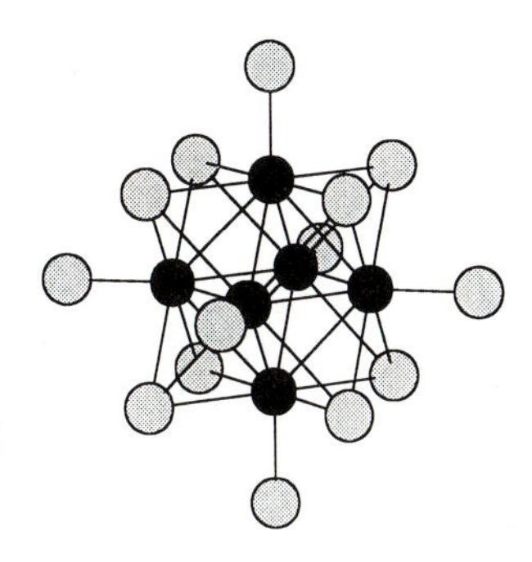

The structure of the $[Mo_6Cl_8Cl_6]^{2-}$ ion

The four metal–metal bonds can be formed in the following ways.

1. *Four single metal–metal bonds.* In the octahedral cluster-anion $[Mo_6Cl_8Cl_6]^{2-}$ each Mo^{II} ion forms four metal–metal bonds along the edges of the octahedron. In the related cluster-ion $[Ta_6Cl_{12}Cl_6]^{4-}$ each metal atom forms four three-centre two-electron bonds on the faces of the octahedron. This leads to a total of eight three-centre two-electron bonds on the eight faces of the octahedron. The structures of the $[Mo_6Cl_{14}]^{2-}$ and $[Ta_6Cl_{18}]^{4-}$ cluster-anions are illustrated in the margin.
2. *Two pairs of double bonds.* In the triangular cluster anion $[Re_3Cl_{12}]^{3-}$ (structure shown in the margin) each Re^{IV} (d^4) ion forms a pair of double bonds to the adjacent metal atoms.
3. *A quadruple bond.* In $[Re_2Cl_8]^{2-}$ the d^4 Re^{IV} ions form a multiple bond of order four, by utilizing the σ, 2π, and δ (d_{xy}–d_{xy} overlap) bonding interactions shown in Fig. M.4.
4. *A triple and a single bond.* The rectangular cluster compound $[Mo_4Cl_8(PEt_3)_4]$ provides an example of this bonding arrangement (see Fig. M.5).

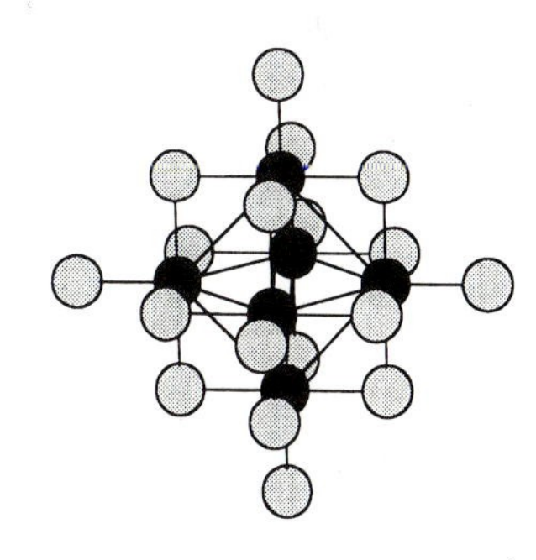

The structure of the $[Ta_6Cl_{12}Cl_6]^{4-}$ ion

In these compounds metal–metal bonding arrangements which have no analogues in main group chemistry are generated, because the metal ions are able to utilize the d_δ metal orbitals in addition to d_σ and d_π.

The early transition metal ions with d^3 electron configurations form triply bonded compounds, e.g. $Mo_2(OR)_6$ and $[W_2Cl_9]^{3-}$, which utilize the d_σ and two d_π metal orbitals.

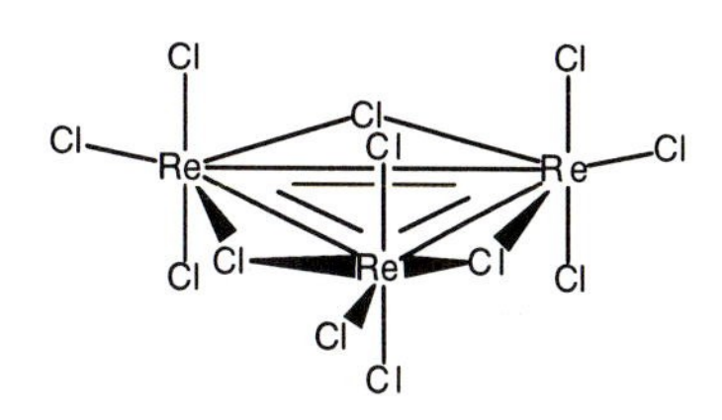

The structure of the $[Re_3Cl_{12}]^{3-}$ ion
Re–Re = 246–247 pm

For complexes of the later transition elements with π-acid ligands the bond order of the metal–metal bonds is determined by the total valence electron count, i.e. the Effective Atomic Number Rule. For a dimeric metal–metal bonded compound this means that 34 valence electrons are associated with a formal metal–metal bond order of 1, 32 valence electrons for a formal bond order of 2, and 30 for a formal bond order of 3. Specific examples of such molecules are illustrated in Fig. M.6. In these compounds there is no formal link between the d electron count and the number of metal–metal bonds formed by the metal atoms.

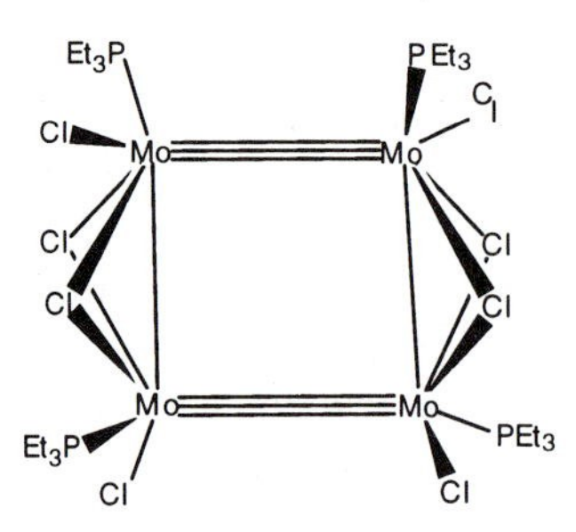

Fig. M.5 The structure of the cluster $[Mo_4Cl_8(PEt_3)_4]$

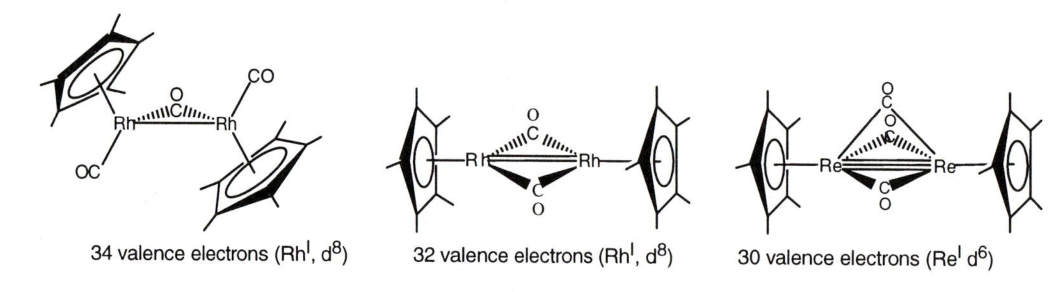

Fig. M.6 Multiple bonding in metal dimers with π-acid ligands

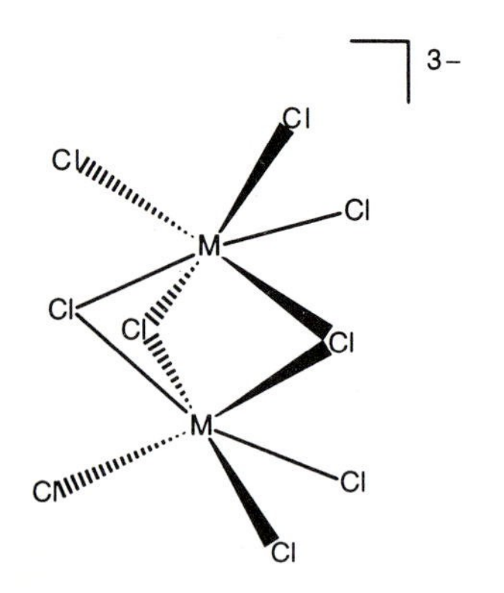

Fig. M.7 Face-sharing bi-octahedral structure of $[M_2Cl_9]^{3-}$

For a more thorough discussion of multiple bonding see F. A. Cotton and R. A. Walton, *Multiple Bonds Between Metal Atoms*, Clarendon Press, Oxford, 1993 and *Structural and Electronic Paradigms in Cluster Chemistry*, Ed. D. M. P. Mingos, Springer-Verlag, Berlin, 1997

Down a column of transition metal atoms the increase in the principal quantum number leads to the maximum in the radial distribution function moving farther from the nucleus and consequently the overlaps increase in the order: 5d–5d > 4d–4d > 3d–3d. This has a profound effect on the relative strengths of the metal-metal bonding.

The $[M_2Cl_9]^{3-}$ (M = Cr, Mo, and W) anions which have structures based on face sharing octahedra as shown in Fig. M.7. The tungsten compound which has the strongest metal–metal bonding is diamagnetic with a metal-metal distance of 242 pm. The metal–metal bond length does not vary greatly if the counter cation is altered. The corresponding molybdenum compounds in contrast show Mo–Mo distances of 252–282 pm depending on the cation and their magnetic properties may be interpreted either in terms of a diamagnetic ground state with low-lying paramagnetic excited states consistent with a molecular orbital scheme where the splittings between the metal bonding and antibonding orbitals are not great or as two d^3 metal ions which are strongly antiferromagnetically coupled. Both these interpretations suggest that the metal–metal bonding is not very strong. In $[Cr_2Cl_9]^{3-}$ the metal–metal distance is much longer (312 pm) and the magnetic properties are consistent with the presence of two Cr^{3+} which are not magnetically coupled. Therefore the structural and magnetic data confirm the absence of metal–metal bonding interactions.

Polyhedral skeletal electron pair theory (PSEPT)

Polyhedral skeletal electron pair theory

The polyhedral skeletal electron pair theory (PSEPT) relates the three dimensional structures of polyhedral molecules to the total number of valence electrons. It represents an analogue of the valence shell electron pair (VSEPR) theory for three-dimensional cage molecules. The simplest three dimensional cage structures have each vertex connected to three other vertices. Some typical three-connected polyhedral skeletal geometries are illustrated in Fig. P.1. Such molecules may be described using conventional two-centre two-electron bonding schemes based on the effective atomic number (EAN) rule.

Since each vertex atom in a three-connected polyhedral molecule, E_nH_n or E_n uses three electrons for forming edge bonds and has either an E–H bond or a lone pair directed away from the centre of the polyhedron. The cluster molecule is therefore associated with a total of $5n$ valence electrons (n is the number of vertex atoms. The main group fragments B–H, C–H, and N–H and main group atoms P, Si, and S are common vertex components and use three hybrid orbitals (approximately sp^3) for bonding to other atoms within the polyhedron and one hybrid either for external bond formation or to accommodate a lone pair as shown below.

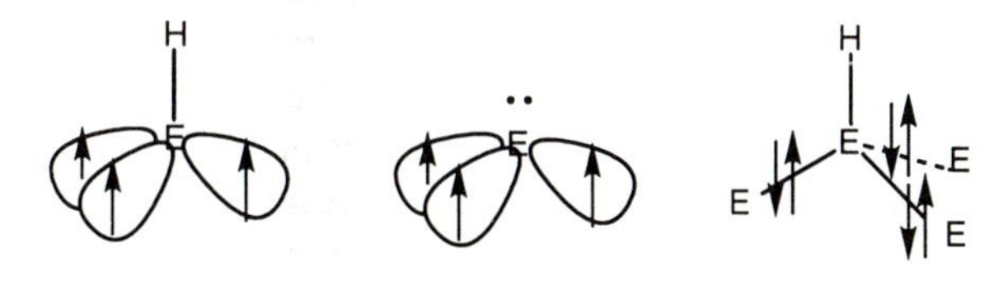

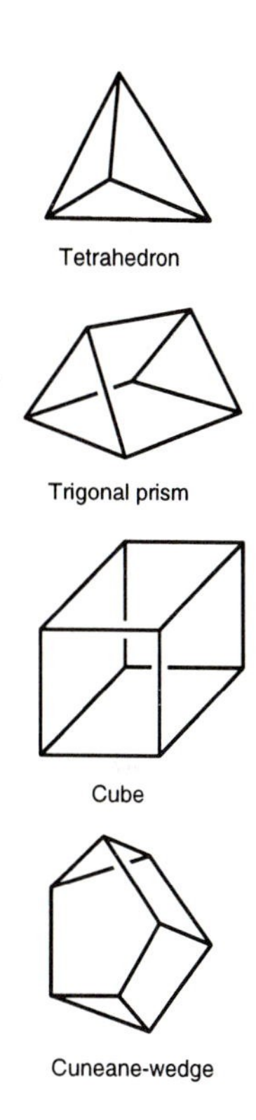

Fig. P.1 Examples of three-connected polyhedra

Some examples of three-connected polyhedral molecules of the main group elements are given in Table P.1.

If the polyhedral skeleton is four-connected it is no longer possible for the E–H fragments to form two-centre two-electron bonds exclusively and therefore a delocalized system of bonding has to be utilized. For example, in octahedral $[B_6H_6]^{2-}$ the delocalized molecular orbitals may be derived by partitioning the three out-pointing hybrid orbitals into radial and tangential components as shown in Fig. P.2.

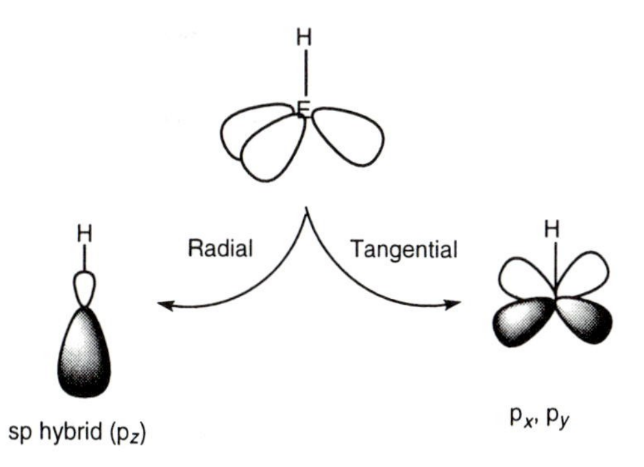

Fig. P.2 The radial and tangential components of the orbitals contributing to the skeletal molecular orbitals of deltahedral molecules

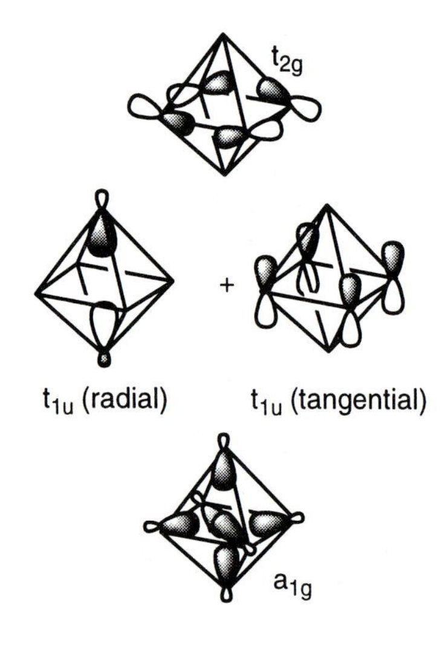

The radial sp hybrids in octahedral $[B_6H_6]^{2-}$ transform as a_{1g}, e_g, and t_{1u}. The a_{1g} linear combination is strongly bonding because all the overlaps are in-phase, the e_g combination is strongly antibonding because the next neighbour overlaps are out-of-phase, but the t_{1u} are non-bonding because there are no contributions from the orbitals on adjacent atoms. The relative energies of the radial molecular orbitals for $[B_6H_6]^{2-}$ are shown in Fig. P.3.

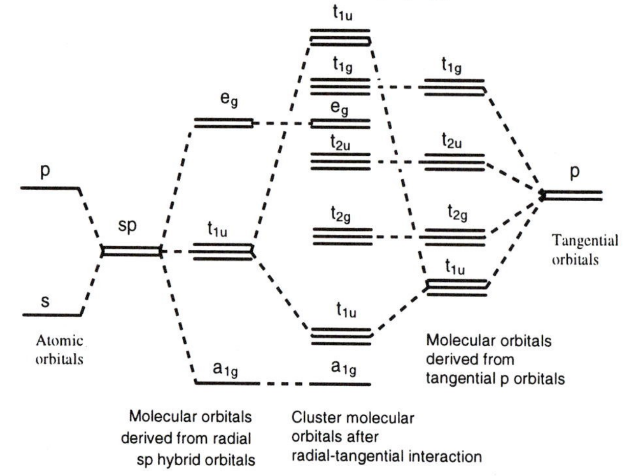

Fig. P.3 The formation of cluster molecular orbitals from the radial sp hybrids pointing into the cluster and the tangential p orbitals in octahedral $[B_6H_6]^{2-}$. The other set of sp hybrid orbitals are used to form terminal bonds between the boron atoms and the hydrogen atoms

Table P.1 Examples of three-connected polyhedral molecules

Geometry	Example	No. of valence electrons ($5n$)
Tetrahedral	P_4	20
	C_4R_4	20
	Si_4^{4-}	20
Trigonal prismatic	C_6R_6	30
	Si_6R_6	30
Cubic	C_8H_8	40
	Si_8R_8	40
	$Si_4O_4R_4$	40
	$Al_4N_4R_8$	40
Cuneane	C_8H_8	40
Dodecahedral	$C_{20}H_{20}$	100

See page 50 for a group-theoretical analysis of this problem

The six pairs of tangential (p_x and p_y) orbitals transform in the octahedral point group as t_{1u}, t_{2g}, t_{1g}, and t_{2u}. The manifold of orbitals is symmetrical about the non-bonding level with six molecular orbitals strongly bonding and six strongly antibonding. If the radial and tangential sub-components of the molecular orbital diagrams are combined (i.e. the t_{1u} radial and t_{1u} tangential orbitals are allowed to mix) it is evident that the bonding in $[B_6H_6]^{2-}$ is characterized by one very stable radial molecular orbital of a_{1g} symmetry and six bonding tangential molecular orbitals of t_{1u} and t_{2g} symmetry.

This makes a total of seven (i.e. $n + 1$) bonding skeletal molecular orbitals. The orbital interaction diagrams for the entirely bonding a_{1g}, the radial/tangential t_{1u} bonding combination, and one of the bonding t_{2g} orbitals are shown in the margin.

The mode of analysis can be extended to other four-connected polyhedral molecules and the conclusions follow a similar pattern, i.e. they all have one very stable radial bonding molecular orbital and n tangential skeletal bonding molecular orbitals. Other examples of four-connected polyhedral skeletons are illustrated in Fig. P.4.

If the electrons participating in E–H bonding are included, the four-connected molecules described above are characterized by a total of $4n + 2$ valence electrons, i.e. $2n$ are associated with the terminal E–H bonds and $2n + 2$ electrons ($n + 1$ electron pairs) occupy the radial and tangential bonding molecular orbitals.

In addition to the four-connected polyhedral molecules described above, the p block elements form a series of cage molecules which adopt polyhedral geometries which have exclusively triangular faces. These polyhedra are described as *deltahedra* (Gk, Δ represents a triangle) as illustrated in Fig. P.5.

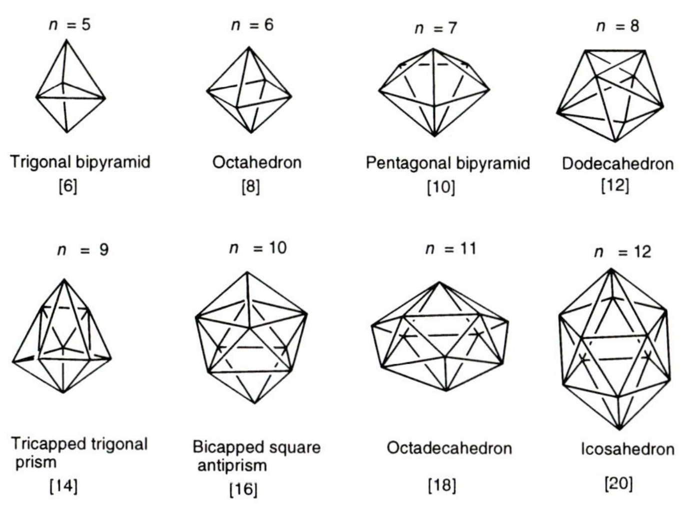

Fig. P.5 Deltahedral skeletons. The numbers in square brackets refer to the number of triangular faces

The vertices of the polyhedra, when projected on a spherical surface, represent the most efficient way of covering the spherical surface and they also have the maximum number of connectivities between the vertices for any spherical polyhedra. The latter property makes them ideal skeletons for electron deficient polyhedra because it maximizes the extent of delocalization by sharing the electrons as much as possible.

The deltahedral molecules E_nH_n illustrated in Fig. P.5 have a pattern of skeletal molecular orbitals which are similar to those for the four-connected polyhedral molecules. They all have $(n + 1)$ skeletal bonding molecular orbitals—1 radial and n tangential. Such molecules are characterized by a total of $4n + 2$ valence electrons—$2n + 2$ occupying the skeletal m.o's and $2n$ occupying the terminal E–H bonds. Examples of deltahedral molecules which satisfy this electronic requirement are summarized in Table P.2. The most important examples of this class of molecule are the borane anions $[B_nH_n]^{2-}$ (n = 6–12) and the isoelectronic carboranes $C_2B_nH_{n+2}$ (n = 3–10). There are also examples of 'naked' deltahedral clusters which have no terminal E–H bonds, e.g. Sn_5^{2-}.

The deltahedral molecules summarized in Table P.2 are described as *closo*-because they have a complete skeleton which can be projected on a spherical shell. In addition there are other series of related polyhedral molecules which have less than complete shells. The *nido*-deltahedra have one of the vertices, usually the most highly connected, removed from the *closo*-structures. Some examples of *nido*-deltahedra are illustrated in Fig. P.6 (the *nido*-description comes from the nest-like appearance of the resultant polyhedral shell, Latin *nidus*—nest).

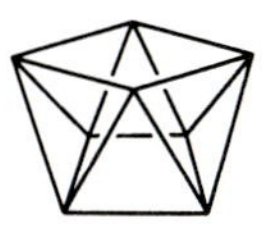

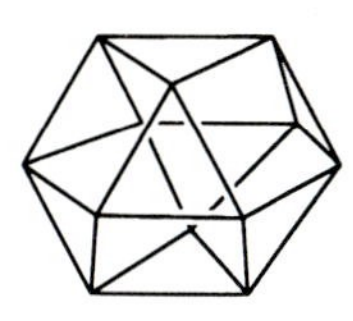

Fig. P.4 Examples of four-connected polyhedral skeletons

Table P.2 Examples of deltahedral cage molecules

Geometry	Example	No. of valence electrons
Trigonal bipyramidal	$[B_5H_5]^{2-}$ $C_2B_3H_5$ Sn_5^{2-} Pb_5^{2-} Tl_5^{7-}	22
Octahedron	$[B_6H_6]^{2-}$ $C_2B_4H_6$ Tl_6^{8-}	26
Pentagonal bipyramid	$[B_7H_7]^{2-}$ $C_2B_5H_7$	30
Dodeca-hedron	$[B_8H_8]^{2-}$ $C_2B_6H_8$	34
Tricapped trigonal prism	$[B_9H_9]^{2-}$ $TlSn_8^{3-}$ $C_2B_7H_9$	38
Bicapped square-antiprism	$[B_{10}H_{10}]^{2-}$ $C_2B_8H_{10}$ Ge_{10}^{2-}	42
Octadeca-hedron	$[B_{11}H_{11}]^{2-}$ $C_2B_9H_{11}$	46
Icosahedron	$[B_{12}H_{12}]^{2-}$ $C_2B_{10}H_{12}$ $[Al_{12}R_{12}]^{2-}$	50

A second series of partial polyhedral skeletons may be derived by removing two vertices from the parent deltahedra. For borane polyhedral molecules two adjacent vertices are removed—the first one used to generate the *nido* plus an adjacent one. Naked main group cage molecules provide examples of cage structures where non-adjacent vertices are removed. Fig. P.6 provides some examples of such polyhedral skeletons, which are described as *arachno* from their resemblance to patterns observed in spiders webs (Greek *arachno* —spider).

A series of *closo*, *nido*, and *arachno*-polyhedral molecules which are based on the same parent have the same number of skeletal bonding electron pairs.

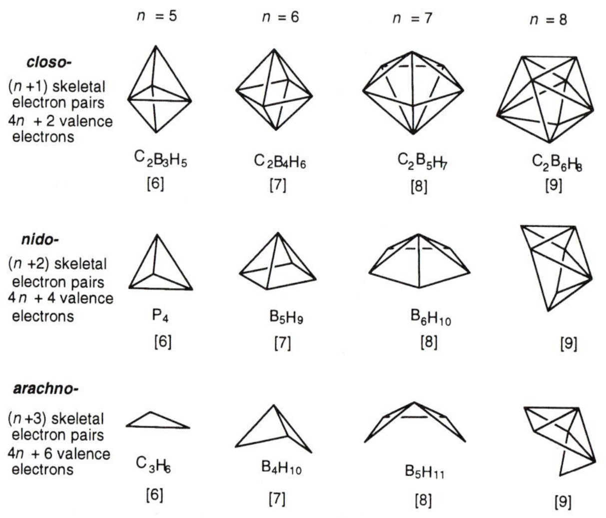

Fig. P.6 Examples of *closo-*, *nido-*, and *arachno*–polyhedral skeletons. The numbers in square brackets refer to the number of skeletal bonding electron pairs

For example, the ions, $[B_6H_6]^{2-}$, $[B_5H_5]^{4-}$, and $[B_4H_4]^{6-}$ can be considered to have their geometries derived from the octahedron. The structures of these ions and their electronic relationships are shown in Fig. P.7.

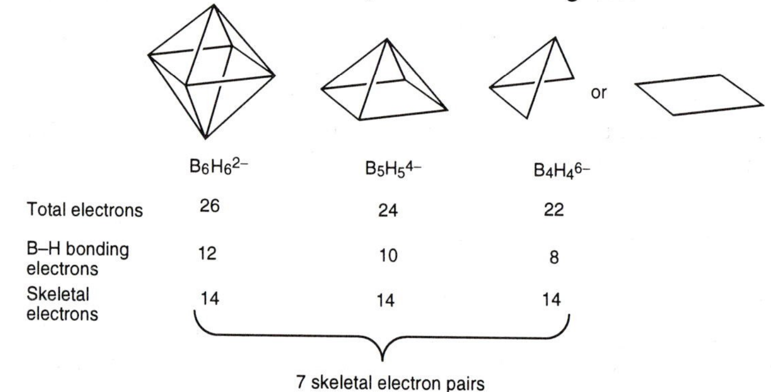

Fig. P.7 Geometric and electronic relationships on *closo-*, *nido-*, and *arachno*–borane

The *arachno*–borane $[B_4H_4]^{6-}$ has two alternative geometries based either on the loss of adjacent or opposite vertices. Each is associated with seven skeletal bonding electron pairs. The borane anions $[B_5H_5]^{4-}$ and $[B_4H_4]^{6-}$ are too basic to be isolated as independent species, but are well documented in their protonated forms B_5H_9 and B_4H_{10}, which retain the same skeletons on protonation. B_4H_{10} has a geometry based on the loss of two adjacent vertices while the Se_4^{2+} ion, which isoelectronic, has a square structure derived from the octahedron by the loss of two opposite vertices.

Transition metal polyhedral molecules

The transition metals form a wide range of compounds in low oxidation states which conform to the effective atomic number rule and the polyhedral molecules which are generated under these conditions also conform to the polyhedral skeletal electron pair theory. However, since the effective atomic number rule for transition metal compounds is based on 18 rather the 8 associated with main group atoms the total number of electrons which characterize specific polyhedral structures are incremented by 10. These relationships are summarized in Table P.3 and illustrated in Fig. P.8.

Table P.3 Total valence electron counts in main group and transition metal cluster compounds, *n* is the number of vertex atoms in the cluster skeleton

Compound class	Total valence electron count	
	for main group example	for transition metal example
Chain	$6n+2$	$16n+2$
Ring	$6n$	$16n$
Three connected polyhedron	$5n$	$15n$
Four connected polyhedron	$4n+2$	$14n+2$
closo-deltahedra	$4n+2$	$14n+2$
nido-deltahedra	$4n+4$	$14n+4$
arachno-deltahedra	$4n+6$	$14n+6$

A wide range of deltahedral boranes and carboranes have been made which are characterized by $4n + 2$ valence electrons. There are analogous examples of deltahedral metal carbonyl clusters with $14n +2$ valence electrons and some examples are illustrated in Fig. P.9. Related examples of *nido*- and *arachno*-metal carbonyl clusters derived from the *closo*-deltahedra by the loss of vertices are illustrated in Fig. P.10.

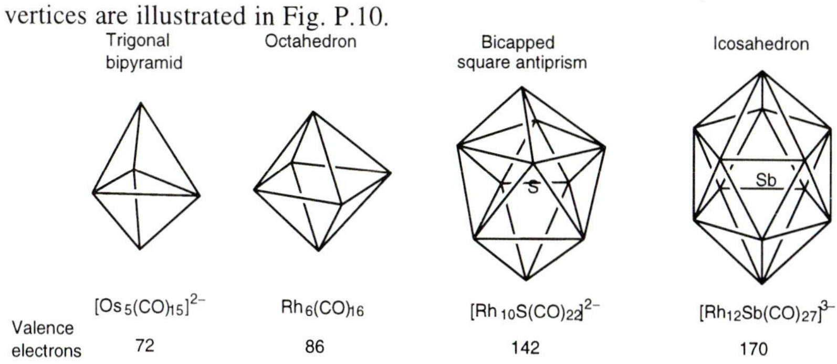

Fig. P.9 Examples of deltahedral metal carbonyl cluster compounds. The locations of the interstitial atoms are indicated

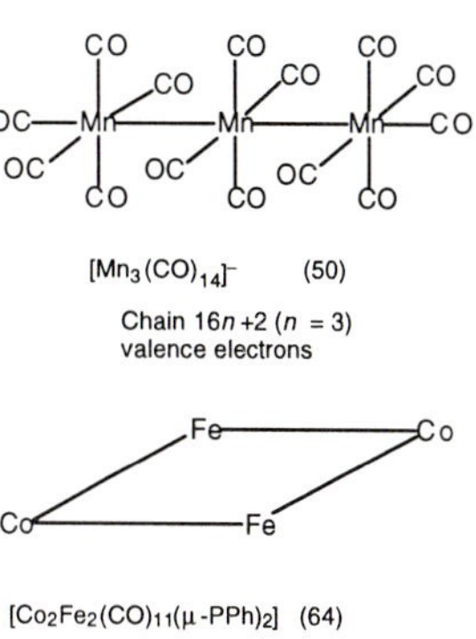

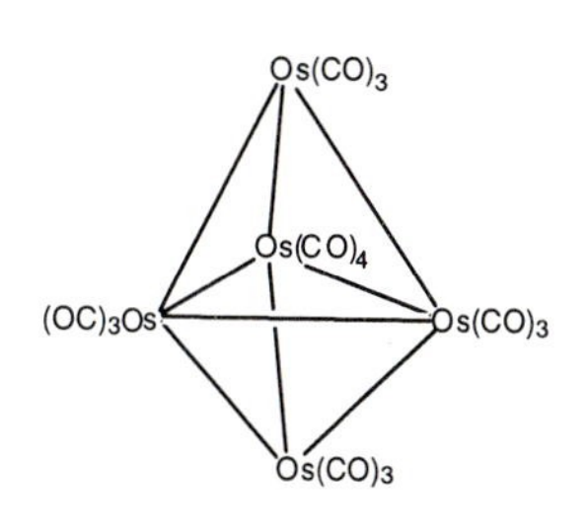

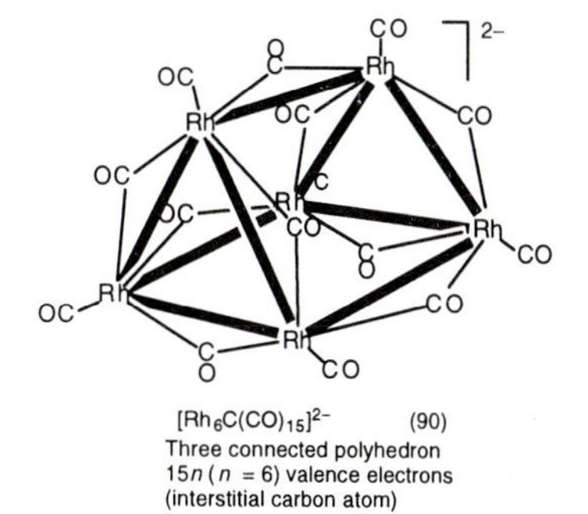

Fig. P.8 Examples of metal clusters which illustrate the entries in Table P.3

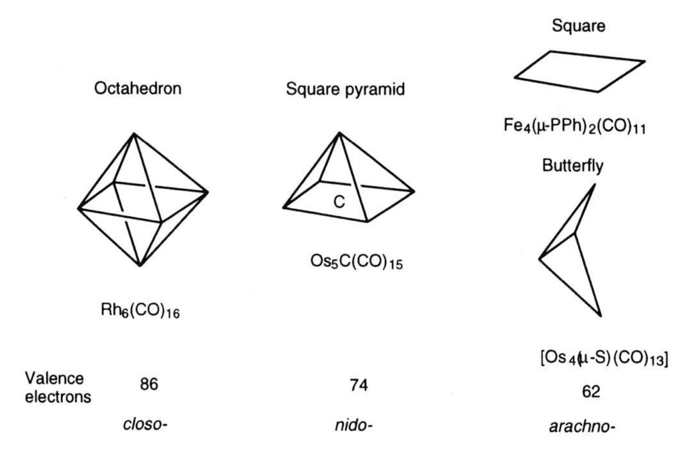

Fig. P.10 Examples of *nido*- and *arachno*-metal carbonyl clusters derived from the *closo*-deltahedra by the loss of vertices

Electron-rich clusters

Main group polyhedral molecules which have between $5n$ and $6n$ valence electrons ($15n$ and $16n$ for their transition metal analogues) have structures which are intermediate between three-connected polyhedra ($5n$) and ring compounds ($6n$). In general, one bond of the three-connected polyhedron is broken for each of the pairs in excess of $5n$. Examples of such molecules are illustrated in Fig. P. 11.

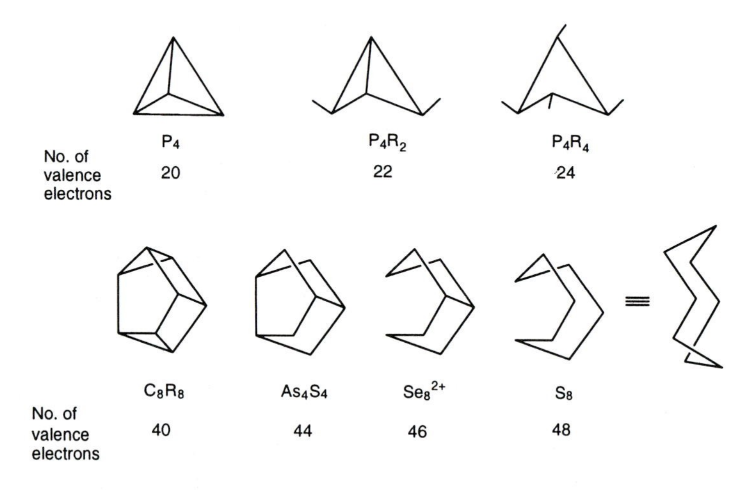

Fig. P.11 Examples of molecules which have structures intermediate between three-connected polyhedra and rings

For a more detailed discussion of these bonding principles see D. M. P. Mingos and D. J. Wales, *Introduction to Cluster Chemistry*, Prentice-Hall, Englewood Cliffs, New Jersey, 1990, and C. E. Housecroft, *Cluster Molecules of the p-Block Elements*, OUP, Oxford, 1994.

Redox reactions

The oxidation–reduction reactions of transition metal complexes proceed via two primary mechanistic pathways; outer and inner sphere reactions.

Outer sphere redox reactions

An outer sphere mechanism occurs when an electron is transferred between two coordination complexes in solution and both of the coordination spheres remaining intact and no ligand is transferred between them. It is only possible to establish such a mechanism unambiguously when both of the coordination complexes are substitutionally inert. The generic mechanism for the redox reaction:

$$[M^{m+}L_6] + [M^{n+}L'_6] \rightarrow [M^{(m+1)+}L_6] + [M^{(n-1)+}L'_6]$$

is illustrated in Fig. R1.

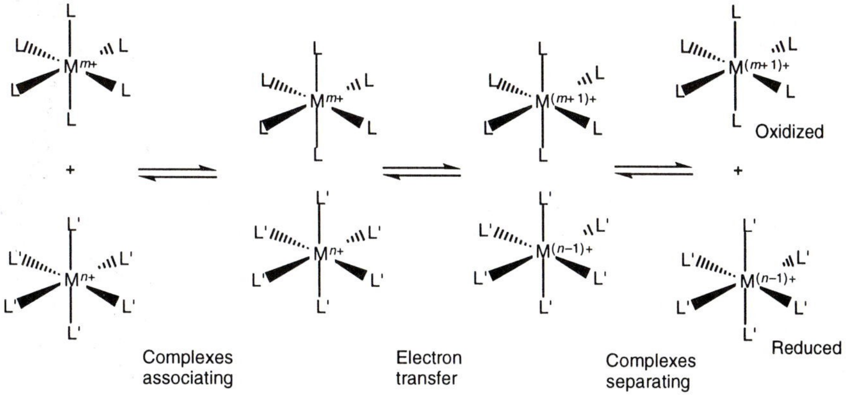

Complexes associating — Electron transfer — Complexes separating — Oxidized — Reduced

Fig. R1 Mechanism for an outer sphere redox reaction

The rate of reaction depends on the energy required to bring together in the transition state the charged complexes and the associated energies required for rearranging their solvation shells (ΔG^*) and the rate of the electron transfer process.

If the electron transfer process is viewed as a tunnelling process, the transmission coefficient for the electron transfer process, κ, appears in the equation for the rate constant:

$$k_{obs} = \kappa Z e^{-\Delta G^*/kT}$$

In the equation κ is the transmission coefficient and Z is the effective collision frequency

Considerable insight has been obtained into this mechanism by studying reactions where the ΔH for the reaction is zero and the thermodynamics of the process is governed by the entropic terms. Examples of such reactions between pairs of compounds which have identical ligands, but different metal oxidation states are:

$$[Fe^{III*}(CN)_6]^{3-} + [Fe^{II}(CN)_6]^{4-} \rightleftarrows [Fe^{II*}(CN)_6]^{4-} + [Fe^{III}(CN)_6]^{3-}$$

where Fe^* represents a radioactive isotope, and

$$\Delta\text{-}[Fe(phen)_3]^{2+} + \lambda\text{-}[Fe(phen)_3]^{3+} \rightleftarrows \Delta\text{-}[Fe(phen)_3]^{3+} + \lambda\text{-}[Fe(phen)_3]^{2+}$$

The redox reaction leads to a racemization which may be followed in a polarimeter.

Such reactions are governed by Franck–Condon restrictions, because the electron transfer process occurs at a much faster rate (10^{-15} s) relative to that required for atomic vibrations (10^{-13} s). This means that the two complexes must have very similar geometries in the transition state. Also, since there is no angular momentum transferred in the transition state the electron transfer cannot be associated with a change in spin state.

For example, the reaction:

$$\underset{t_{2g}^6e_g^1}{[Co^*(phen)_3]^{2+}} + \underset{t_{2g}^6}{[Co(phen)_3]^{3+}} \rightleftarrows \underset{t_{2g}^6}{[Co^*(phen)_3]^{3+}} + \underset{t_{2g}^6e_g^1}{[Co(phen)_3]^{2+}}$$

has a rate constant which is a factor of 10^4 faster than that for:

$$\underset{t_{2g}^5e_g^2}{[Co^*(NH_3)_6]^{2+}} + \underset{t_{2g}^6}{[Co(NH_3)_6]^{3+}} \rightleftarrows \underset{t_{2g}^6}{[Co^*(NH_3)_6]^{3+}} + \underset{t_{2g}^5e_g^2}{[Co(NH_3)_6]^{2+}}$$

This is because in the former case the electron is transferred from the e_g orbital of one complex to the e_g orbital of the second, whereas the second reaction requires a change in spin multiplicity.

In general, octahedral complexes which have electron configurations which involve only changes in the populations of the t_{2g} orbitals tend to have geometries which have closely comparable bond lengths and consequently have low activation energies. However, if they differ in the populations of the e_g orbitals their geometries involve significant differences in bond lengths and the activation energies are high.

The reaction rates given in Table R.1 may be interpreted by employing the consequences of the Franck–Condon restrictions. The data in the table also indicate that the delocalizability of the ligand is also important because the electron density in the metal centre has to be transmitted to the surface of the ligand prior to electron transfer.

Table R.1 Some rate data for outer sphere redox reactions

Reacting pair	Configurations	$k_{25°C}$ /dm^3 mol^{-1} s^{-1}	Orbital correlation	M—L difference / pm
$[Ru(bipy)_3]^{2+/3+}$	t_{2g}^6 / t_{2g}^5	4.2×10^8	$t_{2g} \rightarrow t_{2g}$	V. small
$[Os(bipy)_3]^{2+/3+}$	t_{2g}^6 / t_{2g}^5	2.2×10^7	$t_{2g} \rightarrow t_{2g}$	V. small
$[Fe(bipy)_3]^{2+/3+}$	t_{2g}^6 / t_{2g}^5	3.7×10^6	$t_{2g} \rightarrow t_{2g}$	V. small
$[IrCl_6]^{3-/2-}$	t_{2g}^6 / t_{2g}^5	2.3×10^5	$t_{2g} \rightarrow t_{2g}$	V. small
$[Ru(NH_3)_6]^{2+/3+}$	t_{2g}^6 / t_{2g}^5	4.0×10^3	$t_{2g} \rightarrow t_{2g}$	4
$[Ru(OH_2)_6]^{2+/3+}$	t_{2g}^6 / t_{2g}^5	20.0	$t_{2g} \rightarrow t_{2g}$	9
$[Fe(OH_2)_6]^{2+/3+}$	$t_{2g}^4e_g^2 / t_{2g}^3e_g^2$	4.0	$t_{2g} \rightarrow t_{2g}$	13
$[Co(phen)_3]^{2+/3+}$	$t_{2g}^6e_g^1 / t_{2g}^6$	4.4×10^{-2}	$e_g \rightarrow e_g$	19
$[Co(en)_3]^{2+/3+}$	$t_{2g}^5e_g^2 / t_{2g}^6$	7.7×10^{-5}	$e_g \rightarrow e_g^*$	18
$[Co(NH_3)_6]^{2+/3+}$	$t_{2g}^5e_g^2 / t_{2g}^6$	8.0×10^{-6}	$e_g \rightarrow e_g^*$	18

* Change in spin multiplicity follows transfer

Inner sphere mechanism

In the inner sphere mechanism the electron transfer is mediated by an ambidentate ligand which is capable of bridging the two metal centres in the transition state and is transferred from one complex to the second in the course of the reaction. Ambidentate ligands, X, which are capable of functioning this way include F^-, Cl^-, Br^-, I^-, NCS^-, CN^-, and NC_5H_4Y (where Y is a group capable of acting as a ligand). The generic mechanism for an inner sphere reaction is illustrated schematically in Fig. R.2. The rate determining step may be controlled by the relative activation energies of the three contributing reactions.

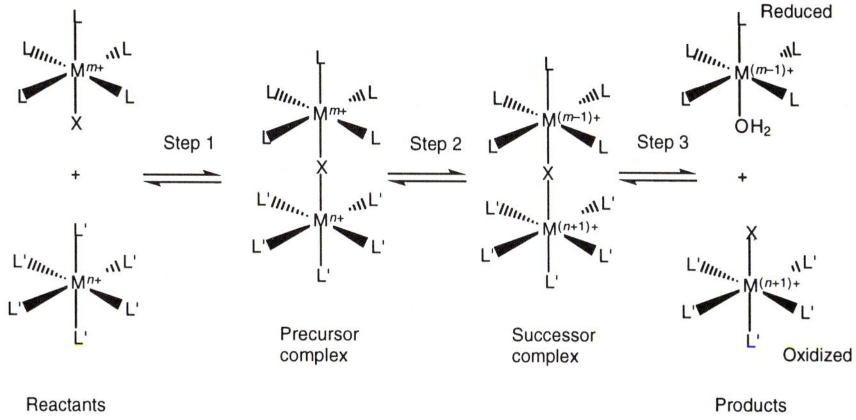

Fig. R.2 Mechanism for an inner sphere redox reaction

Step 1. The rate of formation of the precursor complex. If the metal ions M^{n+} and M^{m+} are substitutionally inert, i.e. they have d^3, d^8, or low spin d^6 electron configurations, then it is difficult for the ambidentate ligand to replace one of the ligands L' in order to form the precursor complex, and consequently this is the rate determining step.

Step 2. If the formation of the precursor complex and the dissociation of the successor complex are rapid then the rate determining step is the electron transfer stage. This rate depends critically on the distance between the metal ions in the transition state.

Step 3. If the complexes formed as a result of the electron transfer step are both substitutionally inert then the dissociation of the successor complex is the rate determining step. For example, if the successor complex is $[(NC)_5Co{-}NC{-}Fe(CN)_5]^{6-}$ then the ions both have d^6 low spin configurations and are substitutionally inert.

Table R.2 gives some rate constants for the reaction of $[Co(NH_3)_5X]^{n+}$ complexes with the $[V(H_2O)_6]^{2+}$ ion and emphasizes how the rate of the inner sphere reaction is influenced by the nucleophilicity of the ligand X. The very low rates for pyridine and NH_3 which are not ambidentate ligands is particularly noteworthy and indicate that they proceed by an alternative outer-sphere mechanism.

Table R.2 Rate constants for the redox reactions between a range of complexes, $[Co(NH_3)_5X]^{n+}$, with the ion $[V(H_2O)_6]^{2+}$. Underlining is used to indicate ligand donor atoms

Ligand, X	k / M^{-1} s^{-1}
I^-	120
Br^-	25
Cl^-	10
$\underline{S}CN^-$	30
$\underline{N}CS^-$	0.3
N_3^-	13
$MeCO_2^-$	1.2
H_2O	0.53
pyridine	4.1×10^{-3}
NH_3	8×10^{-5}

For a fuller discussion of redox reactions and their mechanisms see R. A. Henderson, *The Mechanisms of Reactions at Transition metal sites*, OUP, Oxford, 1993, p 46–60.

S

Stereochemical non-rigidity to synergic bonding

Technique	Interaction time/s
X-ray, neutron, electron diffraction	10^{-18} averaged over unit cells
Microwave spectroscopy	~10^{-10}
Infra-red spectroscopy	10^{-13}
Raman spectroscopy	10^{-14}
Electronic spectroscopy	10^{-15}
N.m.r. spectroscopy	$10^{-1} - 10^{-9}$

All the above techniques of measuring molecular properties except for n.m.r. spectroscopy take times which are too short to detect any intramolecular movements

For a more complete discussion see J. W. Fuller in *Encyclopaedia of Inorganic Chemistry*, Ed., R. B. King, p. 3914, John Wiley and Sons, Chichester, 1994

Stereochemical non-rigidity

Initially one may think that the coordination geometry which is observed in the solid state for a compound is maintained in solution. After all one knows from organic chemistry that the tetrahedral geometry is almost universal for all saturated carbon atoms. However, spectroscopic measurements have indicated that in many inorganic compounds the coordination polyhedron is not rigid and the geometry of the complex varies with time. Whether these changes are observable by spectroscopic techniques depends on the time scale of the experiment relative to the time required for the molecule to rearrange its coordination polyhedron.

The energetics associated with rearranging coordination geometries depends in part on how much the atoms have to move in order to complete the transformation and also on the symmetries of the orbitals involved. For example, tetrahedral carbon compounds are stereochemically rigid because their interconversion would require an intermediate square-planar geometry which necessitates the formation of dsp^2 hybrids rather than the sp^3 hybrids characteristic of the tetrahedron. Since the d orbitals are very high lying this poses problems for molecules based on carbon or other main group atoms. In contrast there are many examples of tetrahedral and square-planar nickel(II) complexes and in solution they interconvert rapidly on the n.m.r. time scale. Ni^{II} ions with a $3d^8 4s^0 4p^0$ ground state has the valence orbitals readily available to form both tetrahedral and square-planar bonding geometries.

Many five, seven, eight, and nine coordinate compounds are stereochemically non-rigid on the nuclear magnetic resonance (n.m.r.) time scale (see the table and note in the margin). For these coordination numbers there are several alternative coordination polyhedra with very similar energies and consequently only minor angular changes are required to interconvert them.

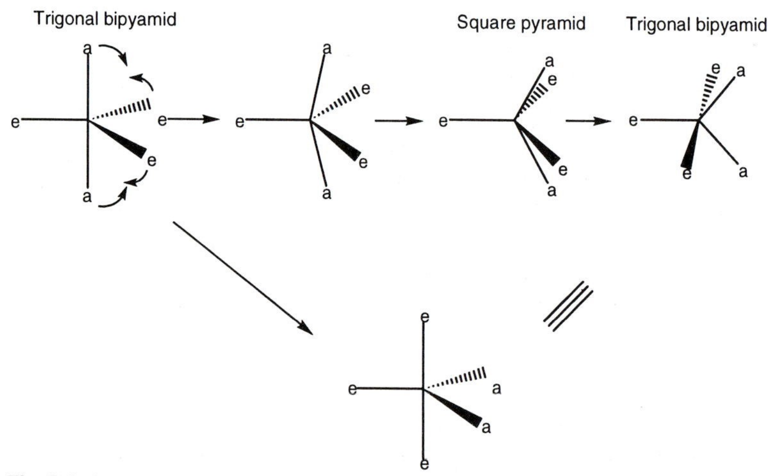

Fig. S.1 A mechanism for the interchange of ligands in a trigonal bipyramidal compound

For example, for a trigonal bipyramidal molecule (D_{3h}) there a two· distinct symmetry inequivalent sites—axial (a) and equatorial (e) in the ratio of 2:3. If the axial and two of the equatorial atoms are moved as shown in Fig. S.1 the a–M–a angle is reduced from 180° and the e–M–e angle is increased from 120°. A point is reached where both of these angles are equal. At this point the coordination geometry corresponds to a square-pyramid (C_{4v}) with the e atom which has remained stationary lying on the four fold symmetry axis.

Although in the parent trigonal bipyramid the a and e atoms were symmetry distinct four of them are symmetry equivalent in the intermediate square pyramid. If the motions of the a and e atoms are continued then eventually a trigonal bipyramid is formed once more, but it has effectively rotated by 90° relative to the initial trigonal bipyramid. This trigonal bipyramid has the axial atoms now occupying equatorial sites and two of the initial equatorial atoms occupying axial sites. Therefore, the simple motions outlined above provide a mechanism for averaging the equatorial and axial environments.

The process is described as a Berry pseudo-rotation because R. S. Berry of the University of Chicago was the first to propose that the internal motions within the molecule described above resulted in what is in effect a 90° rotation of the molecule as a whole. There are many examples of stereochemically non-rigid five-coordinate complexes, particularly in phosphorus(V) compounds and d^8 metal complexes. In some cases, e.g. PF_5 and $Fe(CO)_5$, the activation energy for the rearrangement process is so small that it is not possible to slow down the process sufficiently at low temperatures for the symmetry inequivalent environments to be observed in nuclear magnetic resonance experiments. Where electronic factors lead to a significant energy difference between trigonal bipyramidal and square pyramidal geometries (e.g. for low spin d^6 transition metal complexes, see page 15) the activation energy is significantly larger and at room temperature the complexes may appear to be stereochemically rigid.

In summary, five coordinate complexes are generally stereochemically non-rigid because there are two alternative coordination geometries with similar energies. The interconversion of these polyhedra provides a mechanism for permuting the symmetry inequivalent atoms. The process is intramolecular and may be effectively studied by variable temperature nuclear magnetic resonance measurements if the frequency of permuting the inequivalent nuclei has the same order of magnitude as the difference in resonance frequencies for the symmetry inequivalent nuclei.

For seven, eight, and nine coordination the occurrence of alternative coordination geometries with similar energies leads to many examples of stereochemically non-rigid molecules. The following coordination polyhedra are usually involved:

7	capped octahedron	capped trigonal prism	pentagonal bipyramid
8	dodecahedron	square-antiprism	bicapped trigonal-prism
9	tricapped trigonal prism	capped square-antiprism	

Octahedral complexes with simple ligands are generally stereochemically rigid, because the energy difference between the octahedral and the trigonal prismatic coordination environments is quite large. The repulsions between

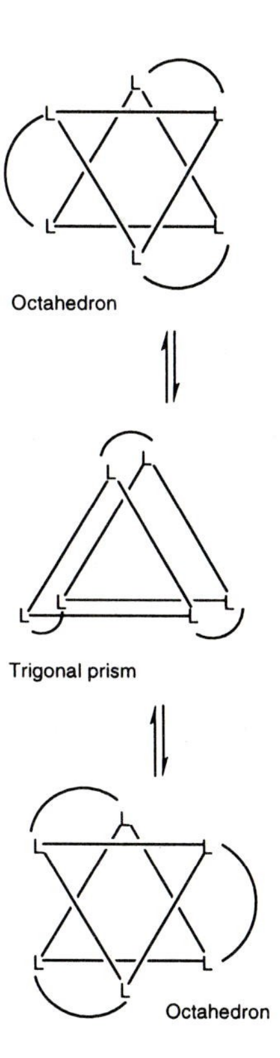

Fig. S.2 The mechanism known as the Bailar twist

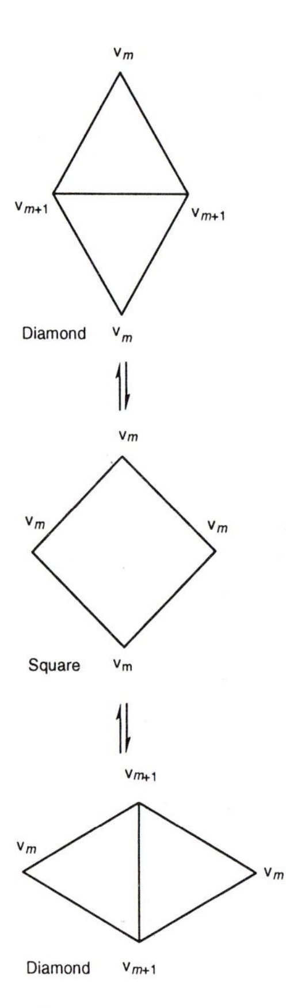

Fig. S.3 The DSD mechanism

the ligands are much larger in the former and this contributes significantly to the energy difference. However, complexes with bidentate ligands where the ligand atoms involved in chelation are forced to remain at approximately the same distance apart rearrange more readily. The process which is illustrated in Fig. S.2 is described as a *Bailar twist.*

It is noteworthy that for the system shown in Fig. S.2 the Bailar twist results in the interconversion of enantiomers. Therefore, if the initial complex is enantiomerically pure and it undergoes a Bailar twist in solution the loss of optical activity provides a means of establishing the rate of the process.

Many of these rearrangement processes are examples of a more general topological process described as a *diamond-square-diamond* (DSD) rearrangement. The process is shown in Fig. S.3. For a polyhedron with two adjacent triangular faces then if the edge common to both triangles is broken and remade using the other two vertices then the process converts one diamond shape into a second rotated by 90° via a square intermediate. This process either converts the initial polyhedron into a different polyhedron or into the same polyhedron with the vertices permuted.

In the process the connectivities (v_m, v_{m+1}) of the vertices which define the diamond change. Those that were initially joined reduce their connectivities by one and the other two increase their connectivities by one. Therefore, if the initial polyhedron has exclusively triangular faces then it is only possible for it to rearrange to an identical polyhedron via a single DSD process if there are pairs of opposite vertices which differ in their connectivities by 1 (see Fig. S.3). Table S.1 summarizes the connectivities of deltahedra, i.e. polyhedra with triangular faces exclusively, and this provides a basis of choosing polyhedra capable of undergoing such rearrangements. The trigonal bipyramid provides such an example since vertices with connectivities of 3 and 4 may be chosen to define the diamond as shown in Fig. S.4. The DSD process illustrated in Fig. S.4 is identical to the Berry pseudo-rotation discussed earlier, but it has been defined in terms of the coordination polyhedron rather than alterations in the metal-ligand bond angles. An alternative diamond for the trigonal bipyramid based on $3v_4$ and v_3 would not lead to a rearranged trigonal bipyramid, but an alternative and indeed unrealizable polyhedron because the connectivities after the rearrangement would be v_2, v_3, and $2v_5$.

The deltahedra of Table S.1 are illustrated on page 63

Table S.1 Connectivities, v_m, of deltahedra with 4–12 vertices

	v_3	v_4	v_5	v_6
Tetrahedron	4			
Trigonal bipyramid	2	3		
Octahedron		6		
Pentagonal bipyramid		5	2	
Dodecahedron		4	4	
Tricapped trigonal prism		3	6	
Bicapped square antiprism		2	8	
Octadecahedron		2	8	1
Icosahedron			12	

For an octahedron the vertices all are four-connected and it is impossible to use a single DSD to generate an equivalent octahedron. A single DSD process results in the conversion of the octahedron ($6v_4$) into a distinctly different polyhedron with $2v_4$, $2v_5$, and $2v_6$—a bicapped tetrahedron (see Fig. S.4). Other deltahedra which are capable of rearranging via a single DSD process are the dodecahedron, the tricapped trigonal prism, and the octadecahedron.

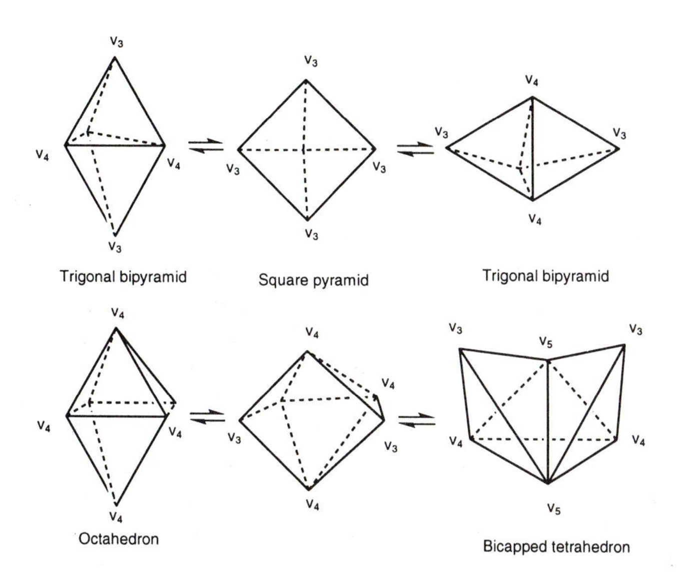

Fig. S.4 The DSD process applied to trigonal bipyramidal and octahedral deltahedra

The polyhedra which are not able to rearrange to an identical polyhedron by a single DSD process require multiple and perhaps simultaneous DSD processes to account for their non-rigidity. For example, the Bailar twist illustrated in Fig. S.2 corresponds to three DSD processes occurring simultaneously.

The discussion may have left the impression that in the solid state all coordination compounds adopt idealised polyhedral structures. However, for those compounds where the potential energy surface connecting the alternative polyhedral geometries is soft then crystal packing effects may favour the adoption of non-regular structures. A careful examination of the structures of a series of related complexes often shows that they have geometries which lie close to the reaction coordinate (the minimum energy path) which interconnects the idealized geometries. These structures therefore give snapshots of the rearrangement process and collectively document the preferred rearrangement pathway.

The presence of a soft potential energy surface may also lead to the observation of alternative coordination geometries for the same ion within one crystal. For example, $[Ni(CN)_5]^{3-}$ has been observed in trigonal bipyramidal and square pyramidal forms within one crystal

Superconductors

In metals the electrical resistance arises either from impurities in the lattice or vibrations of the metal atoms which produce instantaneous irregularities in the otherwise regular three dimensional lattice. Both of these effects impede the smooth movement of electrons through the lattice. The effect of impurities may be reduced by refining the metals and their influence is independent of temperature. The vibrations of the metal atoms are reduced as the temperature is decreased and therefore in general the resistivity of a metal decreases as shown in Fig. S.5. For certain metals the resistivity show a very dramatic and sudden decrease in resistivity at a critical temperature, T_c, near absolute zero. The resistivity suddenly falls to zero as shown in Fig. S.6 and the metal becomes a superconductor. Some typical T_c values for metals and alloys are given in Table S.2.

Table S.2 T_c values for some metals and alloys

Metal/alloy	T_c(K)
Hg	4.2
Sn	3.7
Pb	7.2
V/Si	17
Nb_3Sn	18
Nb_3Ge	23.2

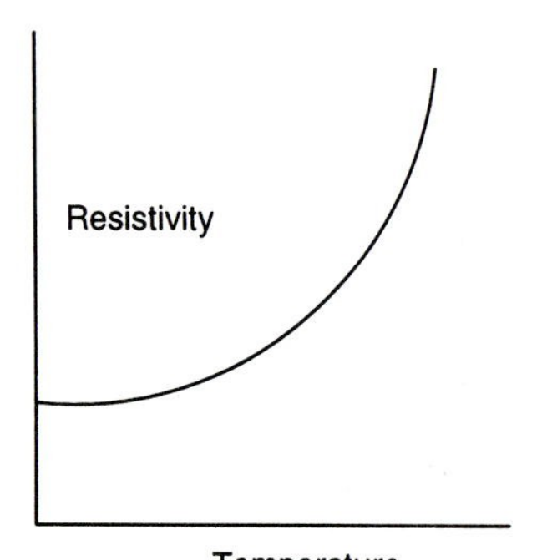

Fig. S.5 Metallic behaviour

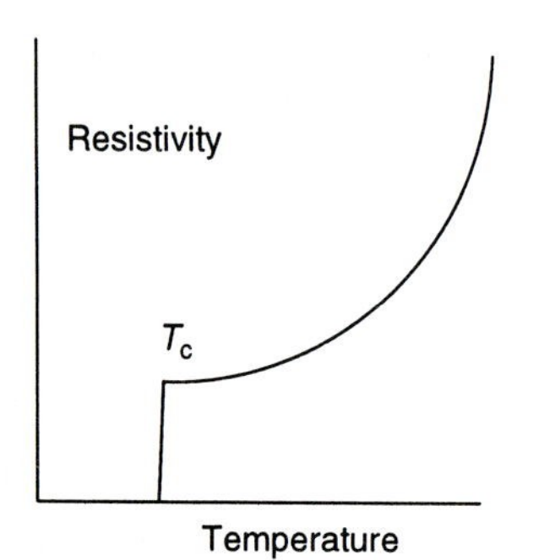

Fig. S.6 Superconductor

The occurrence of superconductivity in metals and alloys is generally interpreted using the BCS (Bardeen, Cooper, and Schieffer) theory which relates the zero resistance to the pairwise motion of the electrons. These electrons pairs are described as Cooper's pairs. The following analogy may be helpful. In motor car rallies which take place on winding roads much higher speeds can be achieved when there are two occupants per car—one driving and one navigating—than those achieved by cars with a single occupant.

At normal temperatures the motion of an electron through a lattice is resisted by the large vibrational motions of the atoms, however as the temperature is reduced these motions have smaller amplitudes and the path of the electron through the more regular maze of atoms is made easier. In these circumstances the location of an electron can itself perturb the lattice. For example, an electron in a bonding region between atoms can make the metal atoms move closer together. When the electron moves on it leaves a cavity in the metal lattice which is very attractive towards a second electron. Therefore the first electron has induced a memory effect which facilitates the motion of the second electron. Therefore, the co-operative movements of the electrons and the nuclei can accelerate the motions of the electrons. No resistance is experienced by the electrons as they move through the lattice if the coupling is really effective. The nuclear motions are easing the flow of electrons in much the same way as a sewing needle is moved through a narrow cylinder of material by a concerted series of squeezing motions.

The average distances between Cooper's pairs of electrons is calculated to be rather large up to 100,000 pm and therefore the interactions are rather weak. Consequently they are only usefully considered as a pair at low temperatures when the vibrational energies associated with the movements of the atoms do not swamp out the effects.

All nuclear particles may be classified as *fermions* or *bosons*. Their angular momentum is related to their nuclear spin. Electrons, protons, and neutrons are fermions, which in general are associated with an angular momentum of $nh/2\pi$ (n is a half-integer multiplier), and are subject to the Pauli exclusion principle. Bosons, e.g. mesons, have an angular momentum of zero or $nh/2\pi$ (where n is an integer multiplier) and do not conform to the exclusion principle

Despite the weak nature of the interaction it is useful to consider the Cooper's pair as a quasi-particle with a charge and mass double those of a single electron. Single electrons are classified as *fermions* and therefore they must conform to the *aufbau principle* and no two electrons can occupy the same quantum mechanical state (Pauli exclusion principle). A Cooper's pair on the other hand is classified as a *boson* and the normal aufbau filling of energy states no longer applies because all Cooper's pairs can occupy the most stable energy state. This provides a mechanism for explaining the very dramatic change in resistivity at T_c. When the energies of the vibrational motions of the atoms become smaller than the energies associated with the formation of Cooper's pairs then the band structure of the metal suddenly changes because instead of many energy levels being filled by pairs of electrons with opposing spins one energy level completely filled with Cooper's pairs is formed. The energy level is quite well separated from the next energy state and therefore the Cooper's pairs move through the lattice like a vast mob ignoring all the scattering nuclei which could cause a resistance to their motion.

The zero resistivity associated with superconductivity is experimentally very difficult to measure and therefore the transition temperature, T_c, is established by direct current magnetic measurements. A superconductor is a perfect diamagnet and persistent currents on the surface of the material expel all magnetic flux. The diamagnetic properties also exhibit some specific hysteresis effects.

Cooling a superconducting sample in the absence of a magnetic field below T_c and then measuring the diamagnetism in an applied magnetic field whilst warming is described as a *zero field experiment* (ZFC). The superconductor exhibits perfect diamagnetism. If the diamagnetism is measured whilst cooling through T_c and then this is followed by measurements on warming they are described as *field cooled* measurements (FC). The magnetic flux present inside the sample above T_c is trapped inside the shielding currents on the surface below T_c. The FC diamagnetic susceptibility is therefore smaller than the ZFC susceptibility. This hysteresis effect is known as the Meissner effect and is the ultimate proof of superconductivity.

In recent years a range of materials have been discovered which are not alloys or metals and yet exhibit high T_c values—in some instances higher than that of boiling liquid nitrogen. Some of these inorganic and organic compounds are summarized in Table S.3. The only two factors they have in common is that they have infinite solid state structures and one of the ions does not have a simple integer oxidation state. The field took an enormous leap forward in 1986 when Bednorz and Muller (IBM, Zurich) reported that a ceramic material based on an lanthanum-barium-copper oxide (see Fig. S.7) became superconducting at 35 K. The discovery of this compound meant that a whole range of substitutional chemistry could be used to introduce variations into the compound's formulae and structures. Indeed, within a year Chu (University of Texas, Houston) had reported that a yttrium-barium-copper oxide had a T_c above the temperature of liquid nitrogen (77.4 K). These 'high temperature ceramic superconductors' therefore display a very dramatic Meissner Effect. If a pellet of the superconductor is placed on a magnet and then the sample and the magnet are cooled by pouring liquid nitrogen over them the pellet is observed to levitate above the magnet when its temperature falls below T_c.

Copper is essential for the observation of superconducting properties in these materials and the introduction of the alternative counter ions has small effects on the copper–copper distances and the structures and these appear to be very important in influencing T_c. The copper–copper distances can also be influenced by putting the ceramic pellets under pressure and generally this has the effect of increasing T_c.

The BCS theory described above which develops the consequences of Cooper's pairs interactions at low temperatures when the atomic vibrations are minimal is clearly not directly applicable to these copper ceramic superconductors. The interaction involving the Cooper's pairs is clearly stronger and this is confirmed by the average distance between the pair of electrons falls to about 3000 pm. It is thought that the anti-ferromagnetic coupling between these copper ions is very important to the formation of the Cooper's pairs in these systems. The oxidation state of copper is crucial in determining whether they are superconducting and their T_c value. In order for the compound to be superconducting the oxidation state has to be either greater than or less than two but not equal to two.

Table S.3 Critical temperatures for some superconducting materials

Compound	T_c / K
$PbMo_6S_8$	13
Ba_6C_{60}	7
Rb_3C_{60}	29
BEDT-TTF[$Cu(NCS)_2$]	10
K_3C_{60}	19
$Nd_{1.85}Ce_{0.15}CuO_4$	22
$Ba_{0.6}K_{0.4}BiO_3$	30
$La_{1.85}Sr_{0.15}CuO_4$	40
$YBa_2Cu_3O_7$	90
$Tl_2Ba_2Ca_2Cu_3O_{10}$	125
$HgBa_2Ca_2Cu_3O_{8+\delta}$	133

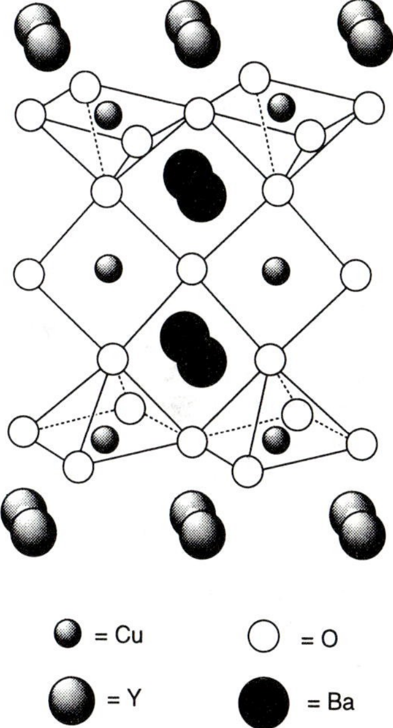

Fig. S.7 The structure of $YBa_2Cu_3O_7$

For example, $LaCuO_4$ is an anti-ferromagnetic insulator, whereas $La_{1.85}Sr_{0.15}CuO_4$ is a metallic conductor and becomes superconducting at 40 K. A wide range of chemical modifications have been introduced into these systems and the T_c has been pushed to 133 K in a mercury derivative (see Table S.3. for T_c values of some inorganic compounds

M. T. Weller, *Inorganic Materials Chemistry*, OUP, Oxford, 1994, gives a detailed discussion of this topic

Table S.3 also includes some other superconducting compounds which have been extensively studied in recent years. Some of them are based on Buckminsterfullerene anions stabilized either by alkali metal or alkaline earth metal cations. These compounds have T_c values between 7 and 29 K. The remainder are organic charge-transfer salts which have T_c values ~10 K. These compounds are clearly not as attractive from an exploitation point of view, but nonetheless have provided some insights into the limitations of the BCS theory.

Symmetry

A chemist intuitively uses symmetry in the process of recognizing which atoms in a molecule are equivalent. For example in SF_5Cl (Fig. S.8), it is easy to see that there are two sets of equivalent fluorine atoms in a ratio 4:1. The recognition of the number of equivalent sites in a molecule is important for analysing the n.m.r. spectrum and determining the number of isomers that may exist.

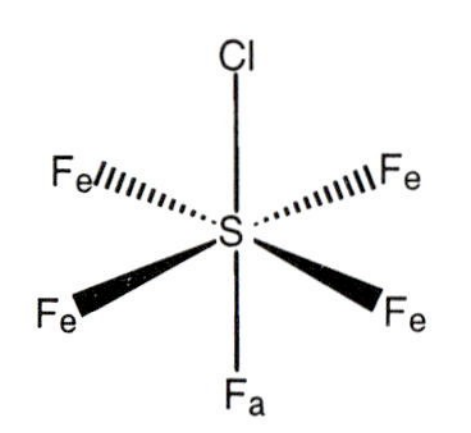

Fig. S.8 The structure of the SF_5Cl molecule. F_e are equatorial fluorine atoms, F_a is an axial fluorine atom

Symmetry plays a particularly important role in the determining the structure of molecules by spectroscopic methods. The interpretation of data from IR spectra, UV spectra, dipole moments, and optical activities all depend upon molecular symmetry.

Symmetry elements and operations

Symmetry is clearly related to the notion of equivalence but a precise definition of symmetry must establish more precisely the mental operations necessary to demonstrate these equivalencies. It is useful to make a clear distinction between symmetry operations and symmetry elements.

A **symmetry operation** is defined as a **movement** *of a molecule towards a new orientation in which every point in the molecule is coincident with an equivalent point (or the same point) of the molecule in its original orientation.*

A symmetry element is a **geometrical entity** such as a **plane**, a **point** or a **line,** with respect to which one or more symmetry operations may be carried out.

Symmetry elements and operations define molecular symmetry and are summarised in Table S.4. Interestingly, the final operation is a combination of two of the earlier symmetry operations and yet represents an independent process. This is therefore analogous to the movement of a knight on a chess board which involves the combination of north-south and east-west movements in a unique way.

Symmetry operations may be illustrated using some specific examples. Fig. S.9 shows the effect of successive symmetry operations on a general point **a**. Rotation by 90° in a clockwise direction about the *z*- axis, which is directed into and perpendicular to the plane of the page, generates a new point **b**, a 180° rotation yields a point **c**, and a 270° rotation point **d.** By definition, these four points must be related by the symmetry operations C_4 (rotation by 90°), C_4^2 (rotation by 2 × 90° = 180°) and C_4^3 (rotation by 270°) about the symmetry axis *z*.

Table S.4 Symmetry elements and their associated symmetry operations

Symmetry Element	Symmetry Operation
1. Centre of symmetry	Inversion of all points through the centre of symmetry (i)
2. Plane	Reflection in the plane(σ)
3. Proper axis	Rotation by an angle ($360^\circ/n$) about the axis (C_n)
4. Improper axis	Rotation by ($360^\circ/n$) followed by reflection in a plane perpendicular to the rotation axis (S_n)

The four equatorial fluorine atoms in SF_5Cl (Fig. S.8) must be related by such symmetry operations and the molecule is therefore said to possess a four fold rotation axis (C_4). Rotation of SF_5Cl about the symmetry axis by 90° (C_4) interchanges the equatorial fluorine atoms and gives an entirely equivalent configuration of SF_5Cl as shown in Fig S.10. Rotations by 180° and 270° (C_4^2, C_4^3) similarly give equivalent configurations. However, rotation by 360°, which could be designated as C_4^4, gives a permutation of fluorine atoms identical with the original configuration, and therefore amounts to doing nothing to the molecule.

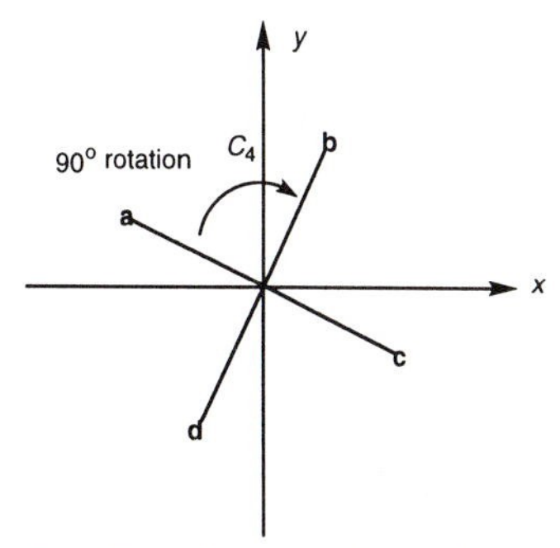

Fig. S.9 An illustration of the effect of a C_4 rotation around the *z* axis on a point **a**

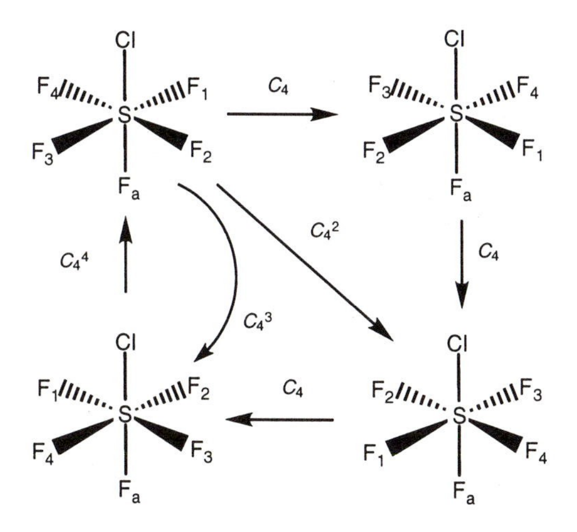

Fig. S.10 The effects of C_4 rotations around the F_a–S–Cl C_4 axis of the SF_5Cl molecule. The four equatorial fluorine atoms are numbered 1–4 to distinguish them as the molecule rotates

Mathematically, the process of doing nothing is itself a symmetry operation, and is called the ***identity operation***, ***E*** (from the German word *Einheit*, meaning unity). The effect of a 360° rotation can therefore be expressed in terms of the following equivalence, $C_4^4 \equiv E$. Quite generally $C_n^n \equiv E$. A molecule may have more than one rotation axis. In such situations, the axis associated with the rotation operation of the highest order is described as the ***principal axis.***

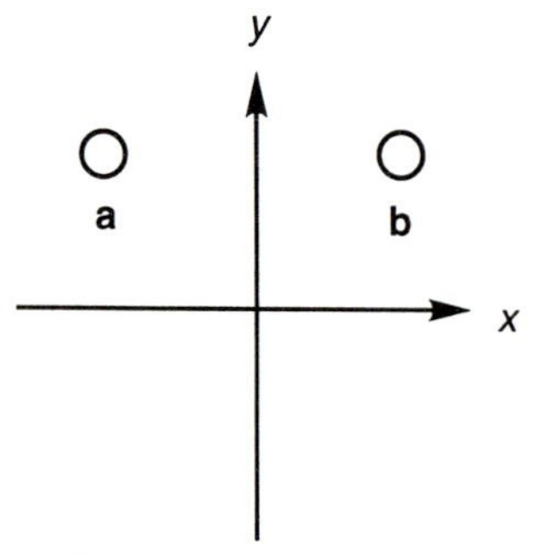

Fig. S.11 The effect of reflecting a point **a** in a reflection plane *yz* is to produce the point **b**

Fig. S.11 shows that when the general point **a** is reflected in the *yz* plane, the point **b** is generated. The oxygen atoms of the sulfur dioxide molecule are clearly related by such a ***reflection operation,*** σ (after the German word *Speigel* meaning mirror). If the mirror plane is perpendicular to the principal axis, it is labelled σ_h (h = horizontal), or if it contains the principal axis σ_v (v = vertical). If the mirror plane contains the principal axis and *bisects* the angle between the two 2-fold axes that are perpendicular to the principal axis, it is labelled σ_d (d = diagonal or dihedral). It is noteworthy that $\sigma^2 = E$ always, i.e. the effect of doing the reflection operation twice returns the molecule to its original permutation of atoms.

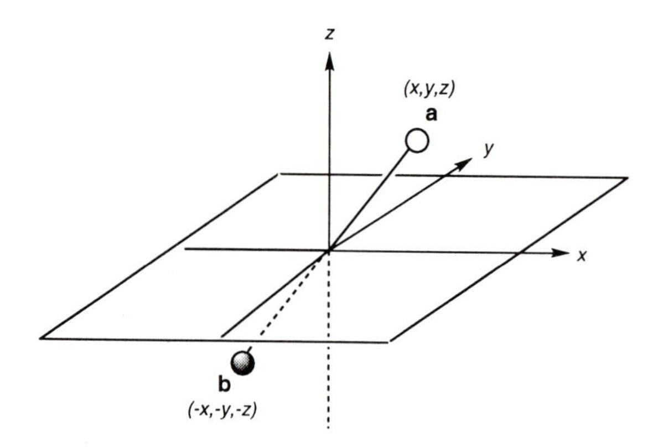

Fig. S.12 An illustration of the inversion operation on the general point **a** producing the point **b**

Fig. S.12 illustrates the effect of an ***inversion operation i*** on the general point **a**, situated at (*x*, *y*, *z*). The new point that is generated by an inversion operation, **b**, has coordinates ($-x$, $-y$, $-z$). In the figure, empty and filled circles have been used to distinguish the points with coordinates *z* and $-z$ (i.e. above and below the plane illustrated) respectively. Note that $i^2 = E$.

The fluorine atoms in *trans*-difluorodiazene (N_2F_2) are related by such a symmetry operation (see Fig. S.13), but not those in *cis*-N_2F_2.

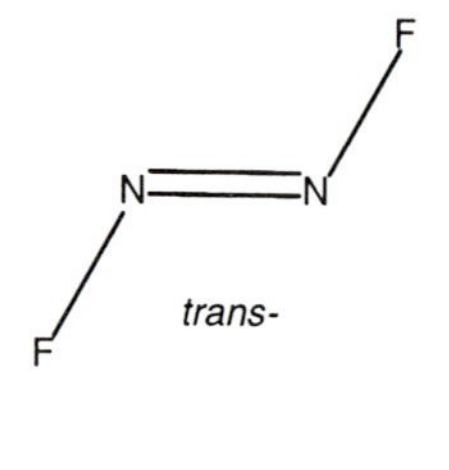

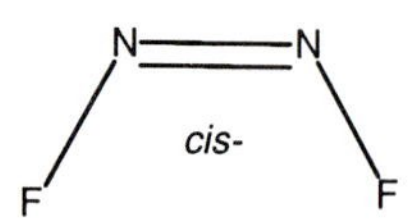

Fig. S.13 The structures of *trans*- and *cis*-N_2F_2

The ***improper rotation operation*** S_n involves a rotation of 360°/*n* followed by a reflection in a mirror plane perpendicular to the rotation axis. Successive improper rotation operations regenerate the original point after a total rotation of 360° *if n is even,* but *if n is odd* two complete revolutions are necessary (i.e. 720°). Fig. S.14 illustrates the equivalent points which are generated by the set of operations S_6, S_6^2, S_6^3, S_6^4 and S_6^4 and Fig. S.15 shows the set resulting from the operations S_3, S_4^2, S_3^3, S_3^4, and S_4^5. The first set of operations generates a staggered arrangement of six points, which would correspond, for example, to the ligand positions in a regular octahedral complex, e.g. $[CrCl_6]^{3-}$ (Fig. S.14), and the second set to an eclipsed arrangement of points. The ligands of a trigonal prismatic complex (Fig. S.15) correspond to such a set.

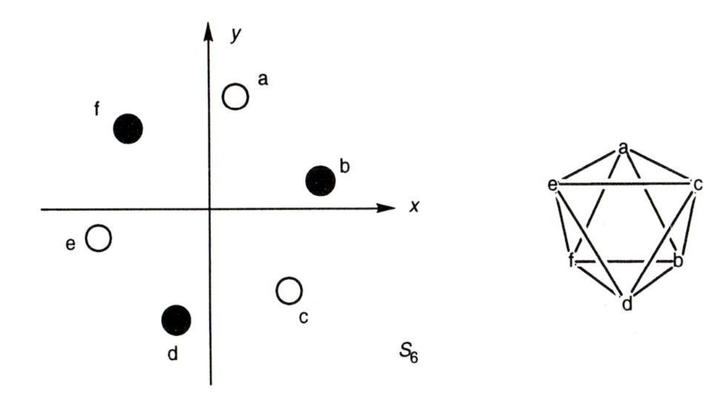

Fig. S.14 Points generated by S_6^m (m = 1–5) operations

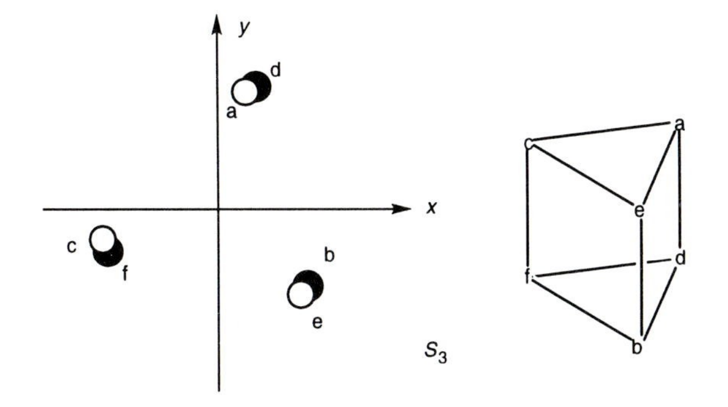

Fig. S.15 Points generated by S_3^m (m = 1–6) operations

If a molecule has a C_n axis and a mirror plane of symmetry perpendicular to it, the C_n axis is also an S_n axis. The application of C_n twice is the same as the application of S_n twice, because the effect of the reflection part of S_n is simply cancelled out, i.e. $S_n^2 = C_n^2$. In general, k applications of S_n will give:

$$S_n^k = \sigma_h \times C_n^k \qquad \text{if } k \text{ is odd, and}$$

$$S_n^k = C_n^k \qquad \text{if } k \text{ is even}$$

Consequently, S_n^k may be interpreted as a rotation of C_n^k followed by a reflection in the horizontal mirror plan only if k is odd. The opposite is also true, i.e. a rotation by $2 \times 2\pi/3$ *plus* reflection if written as S_3^5 and not as S_3^2, which would be simply C_3^2. It follows that:

$$S_1 = \sigma$$

and $S_k^k = \sigma$ if k is odd and $S_k^k = E$ if k is even

Mathematical Group

Mathematically, a **group** is defined as a collection of elements [**P, Q, R,** etc.], with the following properties.

1. There is a rule for 'combining' the elements of the group. When two elements of the group are combined, the result must also be a member of the group, e.g. **P Q = R**
2. There is an element **E**, such that: **E P = P E = P** and **E Q = Q E = Q** etc. **E** is called the identity element.
3. Associative (but not necessarily commutative) multiplication applies i.e. **P (Q R) = (P Q) R**
4. Every element must have an *inverse* that is also an element of the group. The inverse $\mathbf{S} = \mathbf{Q}^{-1}$ of an element **Q** is defined in the following way: **Q S = S Q = E**

Similarity transformations and classes of symmetry operations

A similarity transformation corresponds to the product $\mathbf{P}^{-1}\mathbf{QP} = \mathbf{R}$ and the operation **R** is the similarity transform of **Q** by **P**. **Q** and **R** are described as conjugate. If the consequence of completing similarity transformations on **Q** with all elements of the group including **E** and **Q** itself results in only **P**, **Q**, and **R** they are described as a *class* of symmetry operations. If the point group has *n* elements the number of elements in each class is *n*/*m* where *m* is an integer ($m \leq n$). In the C_{3v} point group (six elements in total) the classes are E; C_3^1 and C_3^2 (two elements, $m = 3$), and $\sigma_v(1)$, $\sigma_v(2)$, and $\sigma_v(3)$ (three elements, $m = 2$)

Multiplication of Symmetry Operations

The consecutive application of two symmetry operations may be represented algebraically by the product of the individual operations. For the sulfur dioxide molecule (Fig. S.16) a proper rotation by 180° about the z axis.

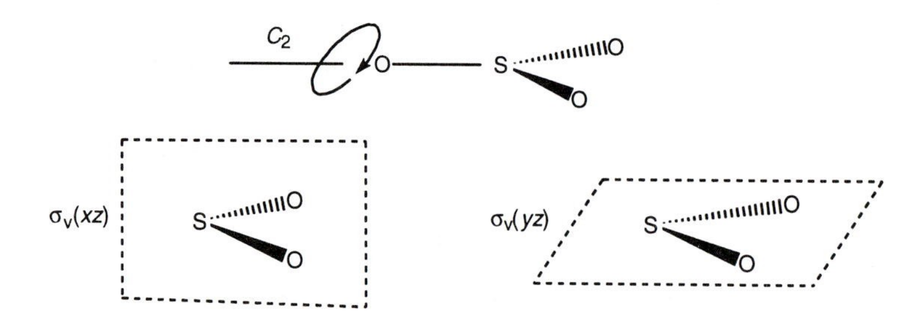

Fig. S.16 The symmetry elements of the SO_2 molecule. The molecule is contained within the *yz* plane

The operation C_2 results in the interchange of the two oxygen atoms; two successive rotations by 180° returns the molecule to its original configuration and so we can state that $C_2 \times C_2 = E$.

The operation of reflection across the mirror plane xz (i.e. $\sigma_v(xz)$) also interchanges the oxygen atoms while reflection in the mirror plane yz (i.e. $\sigma_v(yz)$) leaves the oxygen atom positions unchanged. The application of either reflection operation twice corresponds to the identity operation:

$$\sigma_v(xz) \times \sigma_v(xz) = E$$
$$\sigma_v(yz) \times \sigma_v(yz) = E$$

The consecutive application of $\sigma_v(yz)$ and C_2 has the overall effect of interchanging the oxygen atoms, a result that could alternatively have been achieved by the single operation $\sigma_v(xz)$. Therefore, $\sigma_v(yz) \times C_2 = \sigma_v(xz)$

For this particular molecule, it would have made no difference if the two operations had been applied in the reverse order (the operations are said, therefore, to *commute*), but this is by no means always true. In general, it is necessary to define precisely the order in which symmetry operations are performed. The effect of the application of two successive symmetry operations may be conveniently represented by a **multiplication table** as shown below.

	E	C_2	$\sigma_v(xz)$	$\sigma_v(yz)$
E	E	C_2	$\sigma_v(xz)$	$\sigma_v(yz)$
C_2	C_2	E	$\sigma_v(yz)$	$\sigma_v(xz)$
$\sigma_v(xz)$	$\sigma_v(xz)$	$\sigma_v(yz)$	E	C_2
$\sigma_v(yz)$	$\sigma_v(yz)$	$\sigma_v(xz)$	C_2	E

It is important to note that no new symmetry operations have been generated by these multiplications and, therefore, the four operations E, C_2, $\sigma_v(xz)$, and $\sigma_v(yz)$ constitute a self-contained or complete set. The set therefore constitutes a complete group in a mathematical sense as described in the margin.

Point groups

The complete set of symmetry operations of a *molecule*, constitutes a mathematical group, termed in fact a **point group** because all the symmetry elements intersect at a point within the molecule which is not shifted by any of the symmetry operations. The point group of the SO_2 molecule is denoted C_{2v} according to the Schönflies notation which is commonly used. The point which is not moved by any symmetry operation is the origin is the sulfur atom.

Using a similar reasoning, the symmetry elements and operations of other point groups may be derived, and the appropriate Schönflies symbol assigned. In Table S.5 the *essential* symmetry elements for the various points are listed. The word 'essential' is used since some of the symmetry elements listed in this table for a given point necessarily imply the existence of other operations which are not listed.

Table S.5 The essential symmetry operations associated with the common point groups

Point group	Essential symmetry elements
C_s	Single symmetry plane
C_i	Centre of symmetry
C_n	A single *n*-fold proper axis of symmetry
D_n	A single C_n axis and nC_2 axes perpendicular
C_{nv}	A C_n and *n* vertical mirror planes
C_{nh}	A C_n axis and a horizontal mirror plane
D_{nh}	Same elements as D_n plus a horizontal mirror plane
D_{nd}	Same elements as D_n and *n* dihedral mirror planes
S_n (*n* even)	A single *n*-fold improper rotation axis
T_d	The symmetry operations characteristic of a tetrahedron
O_h	The symmetry operations characteristic of an octahedron or cube
I_h	The symmetry operations characteristic of a regular icosahedron or pentagonal dodecahedron
K_h	The symmetry of a perfect sphere

A systematic procedure for symmetry classification of molecules

The following steps provide a systematic methodology for classifying molecules according to their point groups.

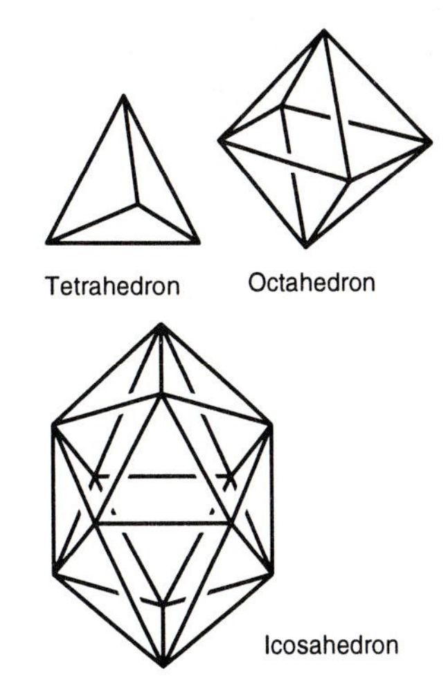

1. Determine whether the molecule belongs to one of the 'special' groups (Table S.5) above), that is $C_{\infty v}$ or $D_{\infty h}$ or to one of those with multiple high-order axes. Only *linear molecules* can belong to $C_{\infty v}$ or $D_{\infty h}$. The especially high symmetry of the others is usually obvious. All of the *cubic groups—T*, T_h, T_d, *O*, and O_h—require four C_3 axes, while the *icosahedral groups*, *I* and I_h require 10 C_3's only and six C_5's. These multiple C_3's and C_5's are the key things to look for. In practice, only molecules built on a central tetrahedron, octahedron, cubeoctahedron, cube, or icosahedron will qualify, and these polyhedra are usually very conspicuous.
2. If the molecule belongs to none of the special groups, search for proper or improper axes of rotation. If no axes of either type can be found, look for a plane or centre of symmetry. If a plane only is found, the group is C_s. If a centre only is found (this is very rare) the group is C_i.

If no symmetry elements at all are evident, the group is the trivial one containing only the identity operation and is designed C_1.

3. If an *even*-order improper axis (in practice only S_4, S_6, and S_8 are common) is found but no planes of symmetry or any proper axis except a collinear one (or more) whose presence is automatically required by the improper axis, the group is S_4, S_6, and S_8.... An S_4-axis requires a C_2-axis; an S_6-axis requires a C_3-axis; and an S_8-axis requires both C_4 and C_2 axes. The important point here is the S_n (*n even*) groups consists exclusively of the operations generated by the S_n-axis. If any additional operation is possible, we are dealing with a D_n, D_{nd}, or D_{nh} type of group. Molecules belonging to the S_n groups are relatively rare, and the conclusion that a molecule belongs to such a group should be checked thoroughly before it is accepted.

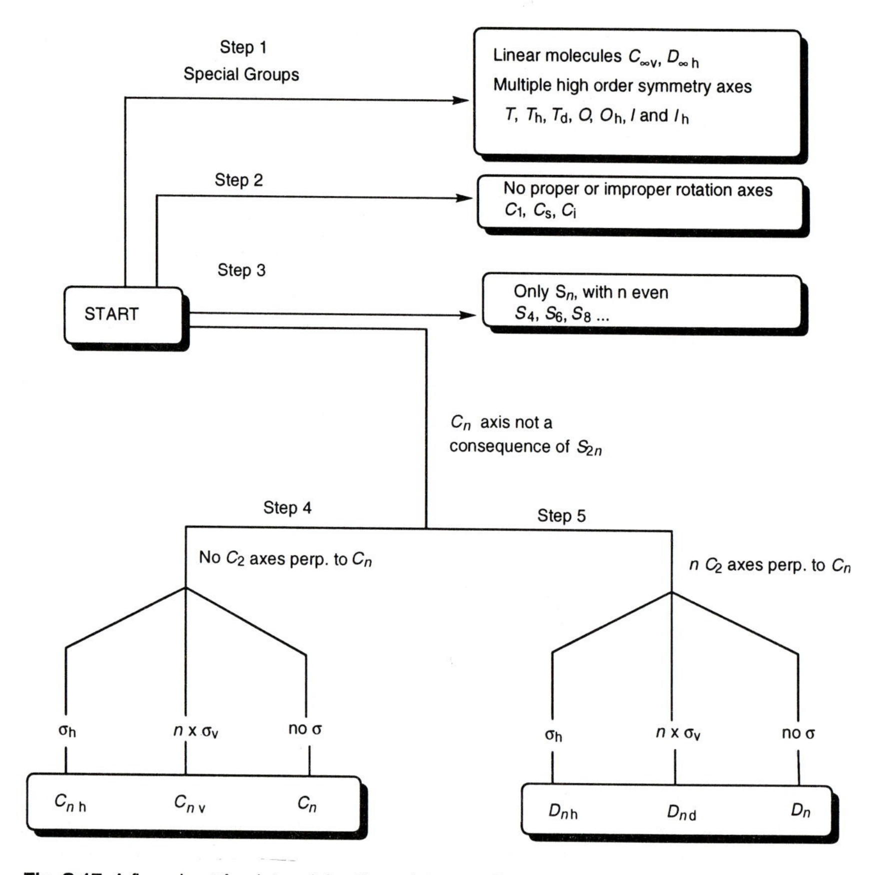

Fig. S.17 A flow sheet for determining the point group for a molecule

4. Once it is certain that the molecule belongs to none of the groups so far considered, look for the highest-order proper axis. It is possible that there will be no axis of uniquely high order but instead some three C_2 axes. In such a case, we look to see whether one of them is geometrically unique in some sense, for example collinear with a unique molecular axis. This occurs with the molecule P_4S_4, which is

one of the examples to be worked through later. If all the axes appear quite similar to one another, then any one may be selected at random as the axis to which the vertical or horizontal character of the planes is referred. Suppose that C_n is our reference or principal axis. The crucial question now is whether there exists a set of nC_2 axes perpendicular to the C_n axis. If so, we proceed to step 5. If not, the molecule belongs to one of the groups C_n, C_{nv}, or C_{nh}. If there are no symmetry elements except the C_n axis, the group is C_n. If there are n vertical planes, the group is C_{nv}. If there is a horizontal plane, the group is C_{nh}.

5. If, in addition to the principal C_n axis, there are nC_2 axes lying in a plane perpendicular to the C_n axis, the molecule belongs to one of the groups D_n, D_{nh}, or D_{nd}. If there are no symmetry elements besides C_n and the nC_2 axes, the group is D_n. Should there also be a horizontal plane of symmetry, the group is D_{nh} group and will also contain necessarily n vertical planes; these planes contain the C_2 axes. If there is no σ_h but there is a set of n vertical planes which pass between the C_2 axes, the group is D_{nd}.

This five-step procedure is summarized in the flow-sheet of Fig. S.17.

Illustrative examples of point group assignment

The scheme just outlined for allocating point groups to molecules will now be illustrated. We shall deal throughout with molecules which do not belong to any of the special groups, and we shall also omit molecules belonging to C_1, C_s, and C_i. Thus, each illustration will begin at step 3, the search for an even-order S_n axis.

Example 1. The molecule SF_4 is set up as shown with the axial fluorine atoms along the x axis and the equatorial fluorine atoms in the yz plane.

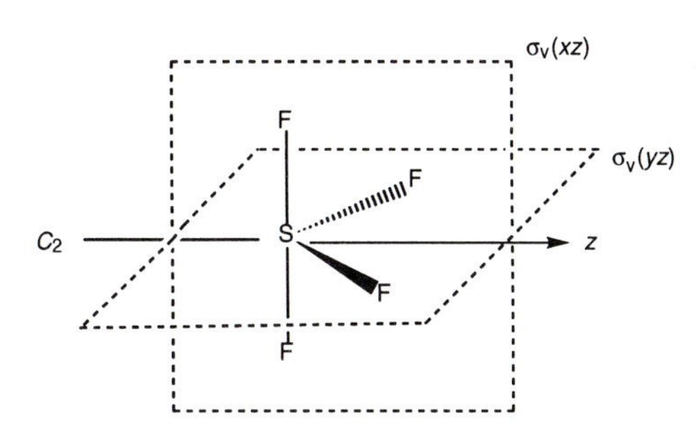

Step 3. SF_4 possesses no improper axis.

Step 4. The highest-order proper axis is a C_2 axis passing through the sulfur atom and bisecting a line between the equatorial fluorine atoms. There are no other C_2 axes. Therefore, SF_4, must belong to C_2, C_{2v}, or C_{2h}. Since it has two vertical planes, one of which is the molecular plane (yz), it belongs to the group C_{2v}.

Example 2. The P_4S_3 molecule is shown in the margin with a C_3 axis passing through the apical phosphorus atom and bisecting the triangle of sulfur atoms.

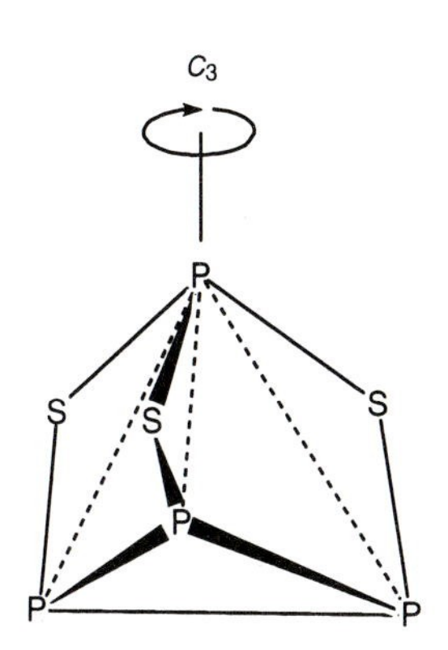

Step 3. There is no improper axis.

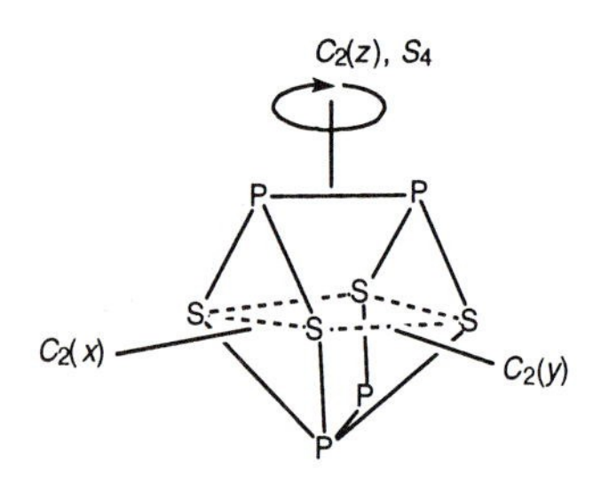

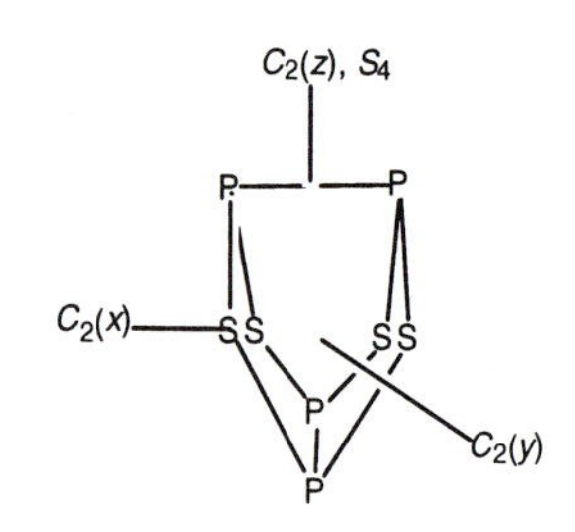

Symmetry equivalent atoms
Atoms in a molecule are symmetry equivalent if they are interchanged as a result of a symmetry operation. In terms of the nuclear magnetic resonance experiment symmetry equivalent atoms have the same chemical shift, i.e. they are isochronous

Magnetically equivalent nuclei
Two nuclei in a molecule are magnetically equivalent if they are symmetrically equivalent by virtue of a symmetry operation that does not simultaneously permute any of the nuclei to which they are spin coupled. An example is shown in Fig. S.19 for the *trans*- and *cis*-forms of the square planar complex, $[(R_3P)_2MH_2]$. In the *trans*-form the ^{31}P nuclei are magnetically equivalent because the operation which permutes them does not permute the hydrogen nuclei to which they are coupled. In the *cis*-form the ^{31}P nuclei are magnetically inequivalent because the operation which permutes them also permutes the hydrogen nuclei

Step 4. The only proper axis is a C_3 axis; there are no C_2 axes at all. Hence, the point group must be C_3, C_{3v}, or C_{3h}. There are three vertical planes, one passing through each P–S–P group. The group is therefore C_{3v}.

Example 3. Two views of the P_4S_4 molecule are shown in the margin.
Step 3. There is an S_4 axis coinciding with the z axis. However, there are also other symmetry elements besides the C_2 axis, which is a necessary consequence of the S_4. Most obvious perhaps, are the planes of symmetry passing through the pairs of adjacent P atoms. Thus, although an S_4 axis is present, the additional symmetry rules out the point group S_4.
Step 4. In addition to the C_2 axis lying along the z axis there are two more C_2 axes perpendicular to this one, as shown in the margin. Thus, the group must be of D-type, and we proceed to step 5.
Step 5. Taking the C_2 axis lying along the z axis of the molecule as the reference axis, we look for a σ_h. One is not present, so the group D_{2h} is eliminated. There are, however, two vertical planes which lie between the C_2 axes, so the group is D_{2d}.

Example 4. The S_2F_2 molecule has a range of rotational isomers depending on the dihedral angle between the two F–S–S planes. The various forms of the S_2F_2 molecule and their C_2 axes (if any) are shown in Fig. S.18.

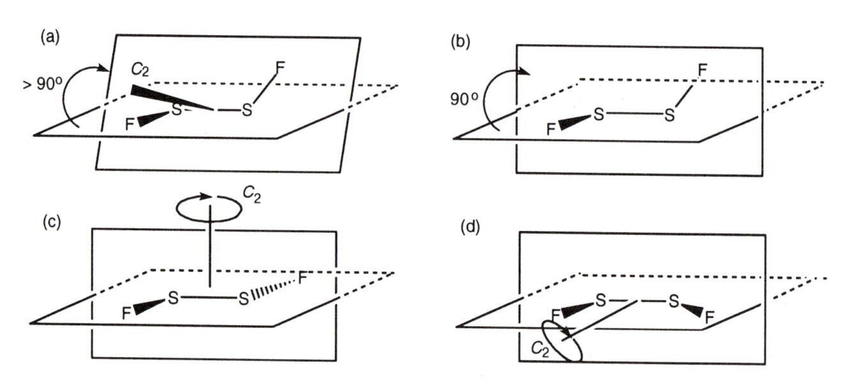

Fig. S.18 (a) The non-planar equilibrium, (b) non-planar with a dihedral angle of 90°, (c) *cis*-planar, and (d) *trans*-planar forms of the S_2F_2 molecule

(i) The non-planar equilibrium configuration ($\theta \neq 90°$), (a) in Fig. S.18.
Step 3. There is no improper axis.
Step 4. As indicated in the figure, there is a C_2 axis and no other proper axis. There are no planes of symmetry. The point group is therefore C_2. Note that the C_2 symmetry is in no way related to the value of angle θ except when θ equals 0° or 90°, in which case the symmetry is higher. We shall next examine the two non-linear equilibrium configurations of the molecule.

The point group remains the same if the dihedral angle is equal to 90° precisely (b) in Fig. S.18, but alternative point groups arise when the atoms all lie in one plane.
(ii) The *cis*-planar configuration, (c) in Fig. S.18.
Again, there is no even-order S_n axis.

Step 4. The C_2 axis, of course, remains. There are still no other proper axes. The molecule now lies in a plane which is a plane of symmetry and there is another plane of symmetry intersecting the molecular plan along the C_2 axis. The group is C_{2v}.
(iii) The *trans*-planar configuration, (d) in Fig. S.18.
Again, there is no even-order S_n axis, (except $S_2 = i$)
Step 4. The C_2 axis is still present and there are no there proper axes. There is now a σ_h which is the molecular plane. The group is C_{2h}.

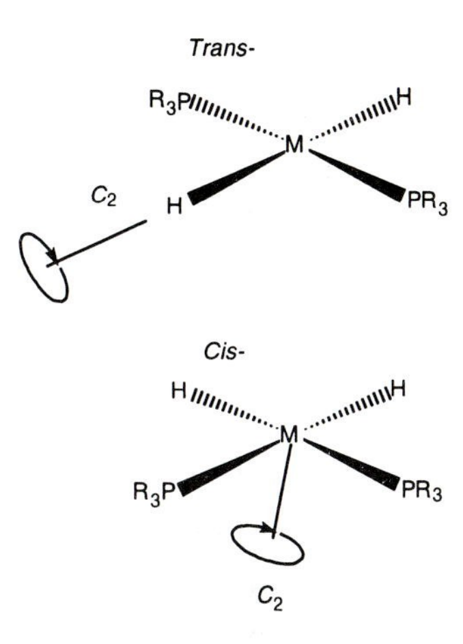

Fig. S.19 *Trans-* and *cis*-forms of $[(R_3P)_2MH_2]$

Symmetry and Optical Activity

Many substances can rotate the plane of polarization of a ray of polarized light. These substances are said to be optically active. In molecular terms, the one necessary and sufficient condition for a substance to exhibit optical activity is that its molecular structure be such that it cannot be superimposed on its image obtained by reflection in a mirror. When this condition is satisfied the molecular exists in two forms, showing equal but opposite optical properties and the two forms are called *enantiomers.*

Whether a molecule is or is not superimposable on its mirror image is a question of symmetry. A molecule that contains a n-fold improper rotation axis of symmetry S_n is always superimposable on its mirror image. This is true because the operation S_n consists of two parts: a rotation C_n and a reflection σ. Since reflection creates the mirror image, the operation S_n is equivalent to rotating the mirror image in space. By definition, a molecule containing an S_n axis is brought into coincidence with itself by the operation S_n and hence its mirror image, after rotation, is superimposable. As $S_1 = \sigma$ and $S_2 = i$, a molecule with either a plane or a centre of symmetry is also optically inactive. However, the most general and economical rule is that *a molecule with an S_n axis is optically inactive.*

Some examples of optically active inorganic molecules are shown in Fig. S.20.

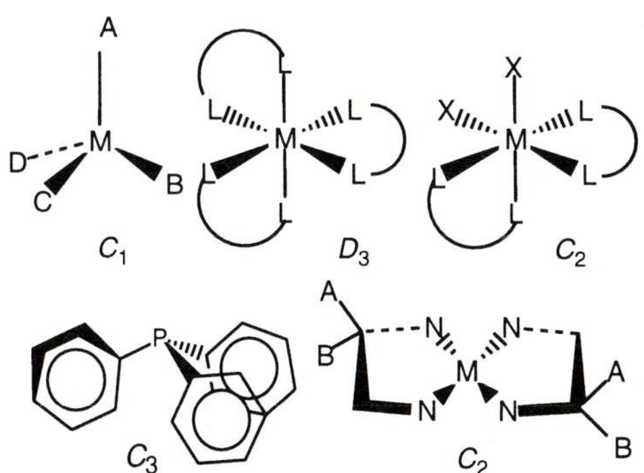

Fig. S.20 Examples of optically active molecules

Chemically equivalent atoms
Two atoms in a molecule are chemically equivalent if one may be replaced by the other in the performance of a proper symmetry operation. Two atoms which are equivalent by virtue of an improper symmetry operation are chemically equivalent only with respect to optically inactive (achiral) reagents

Diastereoisomerism
A ligand (or more generally a molecular fragment) which has two groups related by a plane of symmetry will lose this equivalence if coordinated to a metal complex lacking a plane of symmetry. The groups are diastereoisomeric and have distinguishable chemical shifts. For example, the methyl groups in PMe_2Ph may be distinguishable when the molecule is coordinated in a complex lacking a plane of symmetry

Descent of symmetry

The idealised geometries described above are not always observed in actual molecules and therefore it is helpful to consider the effect of distortions on the point groups of commonly encountered molecules. For example, a regular octahedron may be distorted either along a four fold symmetry axis or a three fold axis. Further more the distortions may be executed in such a way that the centre of symmetry is retained or lost. The resulting point groups are illustrated in Figs S.21 and S.22. Corresponding distortions of the tetrahedron are illustrated in Fig. S.23.

The reduction in symmetry may also be achieved by substituting some of the atoms in the molecule. For example, the introduction of a single substituent into the octahedron leads to the point group C_{4v} which is identical to the point group resulting from lengthening a single bond of the octahedron along the four-fold axis. The introduction of two *trans*-substituents results in a lowering of the symmetry to D_{4h}, whereas the introduction of two *cis*-substituents leads to the C_{2v} point group.

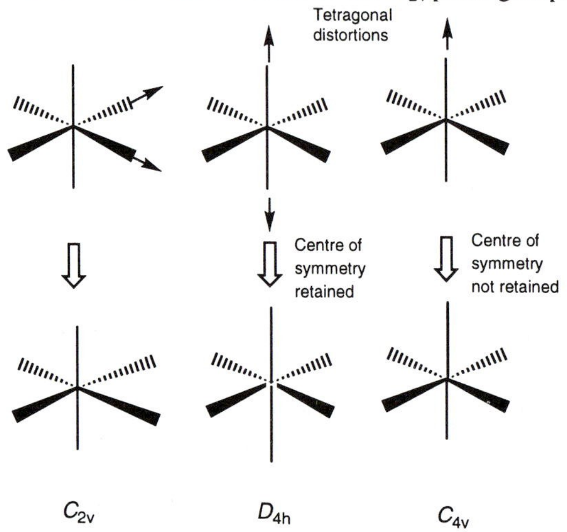

Fig. S.21 Descent of symmetry for the octahedron (distortions along a four-fold axis)

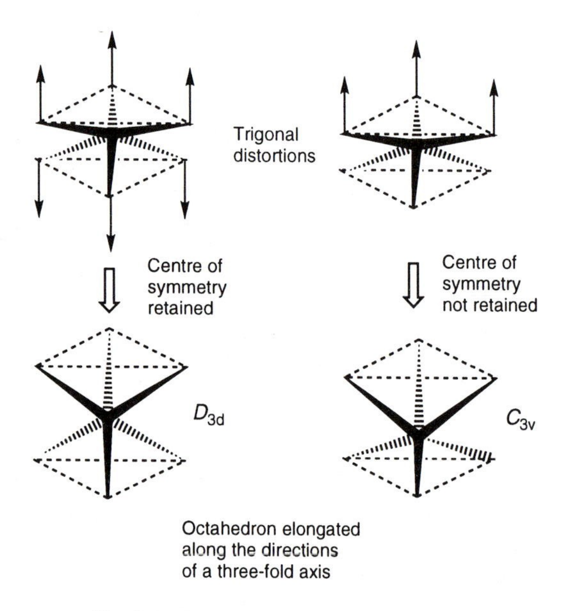

Fig. S.22 Distortions along a three-fold axis

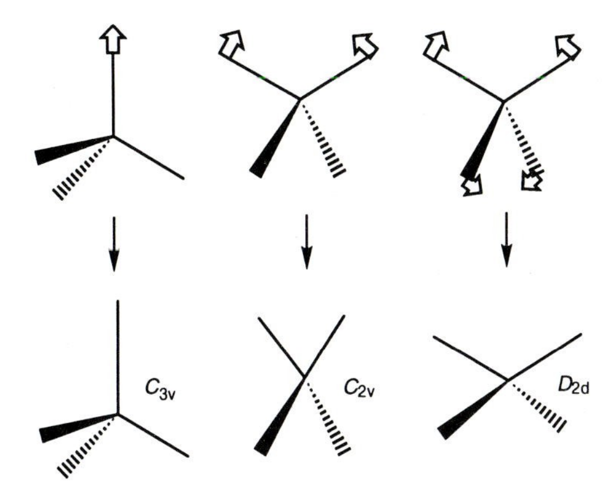

Fig. S.23 Distortions associated with the tetrahedron

Synergic bonding

Ligands such as CO and N_2 form only very weak complexes with conventional Lewis acids such as BF_3 because their lone pair donor orbitals have high ionization energies and are not effective in donating electron density. They nevertheless form stable complexes with transition metals in low oxidation states. This achieved because the σ-donation is supplemented by back donation from filled metal d orbitals to the relatively low lying antibonding π^* orbitals of the ligands. The σ-donation of electron density from the CO ligand to the metal, and back donation from filled d orbitals on the metal to empty π^*-orbitals of the ligand both of which have π-symmetry with respect to the M–C bond are illustrated in Fig. S.24 and a valence bond representation of the redistribution of multiple bonds is illustrated in the margin.

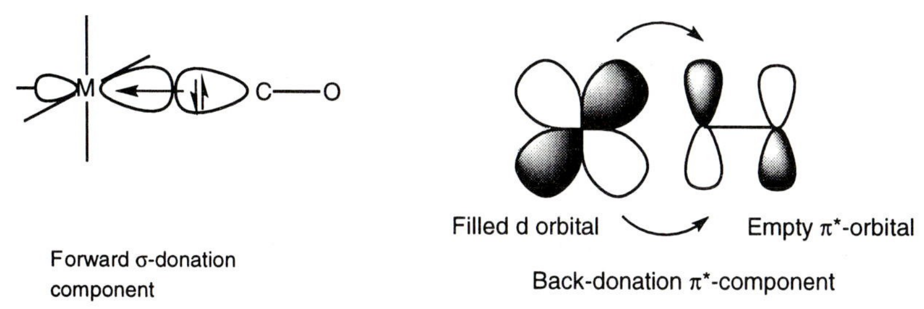

Fig. S.24 A representation of the σ-donation and π-back donation of electron density from a filled metal orbital to the anti-bonding π* orbital of the CO ligand

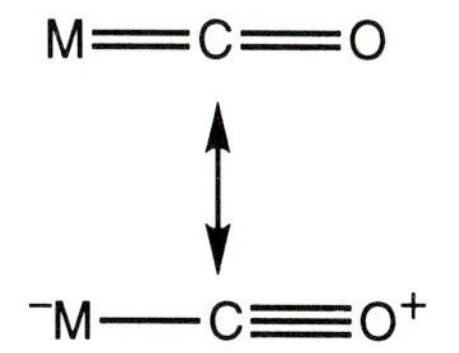

The complementary σ and π components ensure that the metal remains approximately electroneutral. The bonding model is described as synergic because the individual components enhance each other and for example more σ-donation to the metal increases the negative charge on the metal and promotes additional back donation.

Since the back donation occurs from filled metal d orbitals to empty CO π^* orbitals the degree of back donation increases the multiple bonding between metal and carbon at the expense of the C–O multiple bond. These effects may be confirmed experimentally by studying the bond lengths, force constants, or stretching frequencies in isoelectronic carbonyl complexes.

For example, in the series of carbonyl compounds given in Table S.6, the ν(M–C) and ν(C–O) stretching frequencies reflect the way in which the M–C bond strengthens and the C–O bond weakens as the negative charge on the complex increases.

Table S.6 M–C and C–O stretching frequencies (/cm^{-1}) for some metal–carbonyl species

	ν (M–C)	ν (C–O)
CO		2143
$[Mn(CO)_6]^+$	416	2101
$[Cr(CO)_6]$	441	1981
$[Mn(CO)_6]^-$	460	1859
$[Ti(CO)_6]^{2-}$		1750
$[Ni(CO)_4]$		2060
$[Co(CO)_4]^-$		1890
$[Fe(CO)_4]^{2-}$		1790

In situations where the CO is unable to enter into π bonding interactions because it has no filled d orbitals the CO stretching frequency is higher than that in free CO, e.g. Me_3AlCO (2185 cm^{-1}), HCO^+ (2184 cm^{-1}), (η-$C_5Me_5)_2Ca(CO)$ (2158 cm^{-1}), H. H. Brintzinger *et al., Angew. Chem. Int. Ed.*, 1995, **34**, 791

As the negative charge is increased the dominant effect involves the back donation component which increases and strengthens the M–C bond at the expense of the CO bond.

Other ligands which are capable of forming multiple bonds of this type include NO, CNR, CS, CR_2 (carbenes), and CR (carbynes). Their relative abilities to act as π acceptors is :

$$NO^+ > CS > CO > CNR > N_2 > CN^-$$

and their σ donating abilities depend on the charge on the ligand and the availability of the electrons in the σ-donor orbital. For example, CN^- is a good σ-donor and poor π-acceptor and found generally in complexes with intermediate oxidation states whereas the other ligands are particularly useful for stabilizing transition metals in low oxidation states.

If the π-acceptor ligand has only one acceptor component then it aligns itself in the complex in such a way that the back donation effect is maximized. Carbenes represent examples of such ligands and the electronic reasons for the adoption of a particular conformation for the compound Ta(η-$C_5H_5)_2CH_3CH_2$ are illustrated in Fig. S.25.

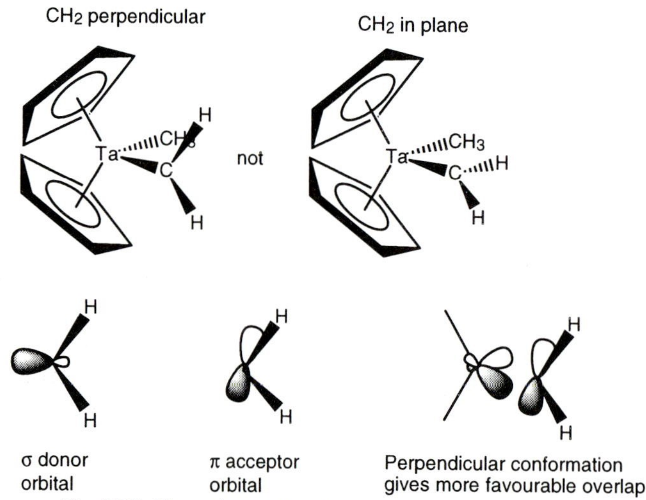

Fig. S.25 Alternative rotational isomers in a carbene complex

Phosphines are also capable of acting as π-acceptor ligands although there has been some controversy regarding the precise mechanism. Initially it was proposed that the phosphines are able to function as π-acceptors through their empty 3d orbitals as shown on the left hand side of Fig. S.26. However, these orbitals are at very high energies and the more favoured current explanation is that they act as π-acceptors through the antibonding P–X molecular orbitals as shown on the right hand side of Fig. S.26.

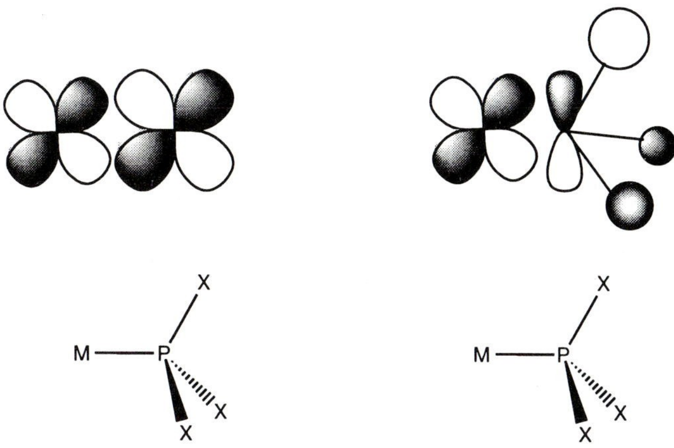

Fig S.26 Back donation from filled metal d orbitals may involve either 3d orbitals on phosphorus (left) or antibonding σ^* (P–X) orbitals (right)

The symmetries of the orbitals involved are the same and therefore there are few simple experiments which can discriminate between the relative contributions of the alternative interactions. Whatever, the precise mechanism the following relative orders of donor and acceptor abilities may be proposed:

σ-donor ability:

$$PMe_3 > PAr_3 > P(OMe)_3 > P(OPh)_3 > P(NMe_2)_3 > PCl_3 > PF_3$$

π–acceptor ability:

$$PF_3 > PCl_3 > P(NMe_2)_3 > P(OPh)_3 > P(OMe)_3 > PAr_3 > PMe_3$$

Taking the two effects as a whole PF_3 ends up as a ligand with properties very similar to those of CO and can be viewed as a very good π-acceptor ligand which can stabilise low oxidation states whereas PMe_3 is a very poor π-acceptor and should be viewed primarily as a good σ-donor ligand. PMe_3 is thus found in complexes where the metal oxidation state may vary from zero to six.

T

Tolman cone angle

Tolman cone angle

The properties of ligands are determined not only by their electronic properties but also their steric requirements. The Tolman cone angle concept represents an attempt to put these steric effects on a more quantitative basis. For complexes of organophosphine ligands the space occupied by the ligand is estimated on the basis of the imaginary cone which radiates from the metal and just touches the Van der Waals surfaces of the hydrogen atoms on the surface of the ligand as shown in Fig. T.1. The methodology may be extended to other commonly used ligands and the relevant data are given in Table T.1. The substituents in a phosphine ligand do not define a perfect cone and the Tolman angle depends on the relative orientations of the rings or alkyl groups and therefore may vary from complex to complex.

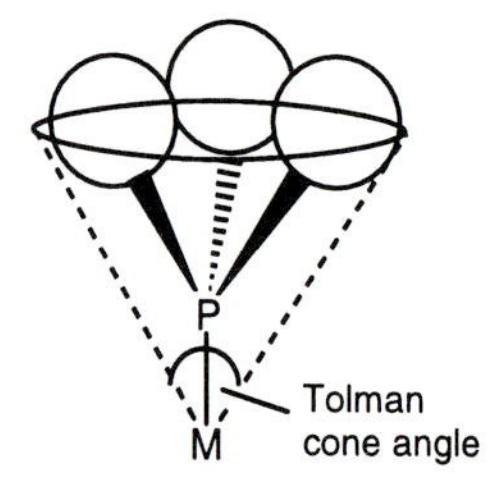

Fig. T.1 An illustration of the Tolman cone angle and its dependence upon the van der Waals radii of the ligands

Since the organic substituents on phosphine ligands change the electronic properties of the ligands and also influence their steric requirements attempts have been made to separate these effects. The pK_a values of the corresponding conjugate acids $[HPR_3]^+$ provide an estimate of the donor properties of the ligands, and the Tolman cone angle estimates the steric requirements of the ligands.

It is significant to note from the data in Table T.2 that the cone angle for phosphine ligands may be doubled by varying the organic group from (OMe) 107° to (2,4,6-$Me_3C_6H_2$) 212° and the pK_a values from –2.00 for $P(OPh)_3$ to 11.40 (PBu^t_3). These variations have enabled chemists to modify in a more controlled fashion the reactivities of metal complexes.

For example, the dissociation of phosphine ligands in $Ni(PR_3)_4$ appear to be dominated by steric effects and the dissociation constant K defined below:

$$Ni(PR_3)_4 \overset{K}{\rightleftharpoons} Ni(PR_3)_3 + PR_3$$

increases in the order:

$$P(OEt)_3 < P(O\text{–}p\text{-tolyl})_3 < P(OPr^i)_3 < P(O\text{–}o\text{-tolyl})_3 < PPh_3$$

suggesting that steric effects predominate.

The relative tendency of Group 15 ligands to replace amines has been estimated by measuring K for the following reaction:

$$W(CO)_5(\text{amine}) + L \overset{K}{\rightleftharpoons} W(CO)_5L + \text{amine}$$

K increases in the order:

$$Ph_3Bi < (PhO)_3P \sim Ph_3Sb < Ph_3As < Ph_3P < (BuO)_3P < PCy_3 < PBu_3$$

This suggests the stabilities of the phosphine complexes are favoured by small ligands with good donor substituents.

Sterically demanding ligands can be used to promote the following properties.

1. *Low coordination numbers.* The ligands $N(SiMe_3)_2$ and PCy_3 are particularly effective at promoting low coordination numbers and some relevant examples are shown in Fig. T.2.

Table T.1 Tolman cone angles (°) for some common ligands

Ligand	Cone angle
H	75
F	92
Cl	102
Br	105
I	107
Me	90
Et	102
Pr^i	114
Bu^t	126
C_5H_5	136
CO	95

2. *Multiple bonds.* Large substituents protect the multiple bond from nucleophiles and thereby make the compound more inert than those with smaller ligands. The reaction shown in Fig. T.3 occurs readily, but this is not the case with the corresponding reaction for the related unsubstituted cyclopentadienyl compound.

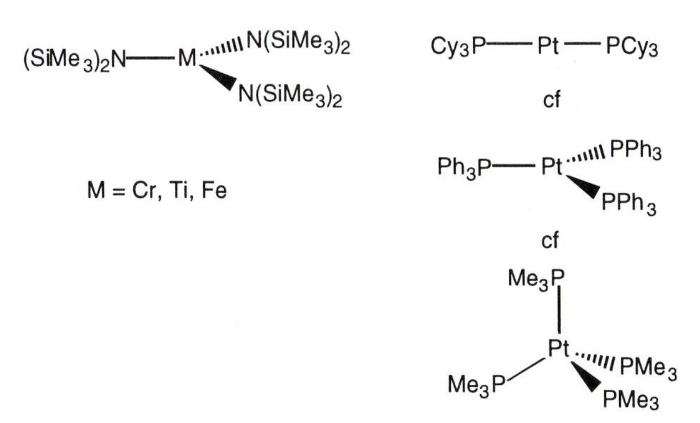

Fig. T.2 Some complexes with low coordination numbers

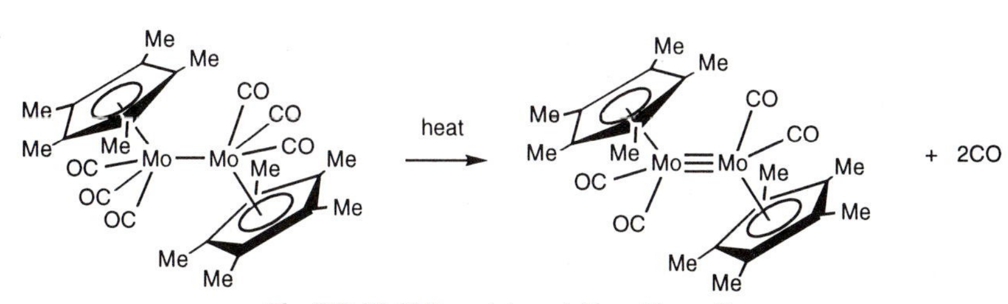

Fig. T.3 Multiple metal–metal bond formation

Table T.2 Tolman cone angles (°) for phosphine and amine ligands

Phosphine/amine	Cone angle	pK_a
$P(OMe)_3$	107	2.6
PMe_3	118	8.65
$P(OPh)_3$	128	–2.00
$P(OPr^i)_3$	130	4.08
PEt_3	132	8.69
PBu^n_3	132	8.43
PPh_3	145	2.73
$P(p\text{-}MeC_6H_4)_3$	145	3.84
PPr^i_3	160	–
PCy_3	170	9.70
PBu^t_3	182	11.40
$P(2,4,6\text{-}Me_3C_6H_2)_3$	212	7.3
NH_3	94	16.46
NH_2Ph	111	10.56
NH_2Bu^t	123	18.14
NMe_3	132	17.61
NEt_3	150	18.46
NPh_3	166	–

The sterically demanding alkoxide ligands, OCH_2SiMe_3, and OBu^t, have similarly been used to stabilize compounds such as those shown in the margin.

RO, RO, RO — Mo≡Mo — OR, OR, OR

R = CH_2SiMe_3, Bu^t

3. *Protecting reactive ligands on the metal centre.* The phosphido-ligand P^{3-} has recently been stabilized in the protected environments provided by $NPhPr^i$ ligands. Similarly, the dihydrogen complex was first isolated in a molybdenum compound where it is protected by the PCy_3 ligands (see compounds below).
4. *Stabilizing unusual oxidation states.* Benzene rings with large substituents have been used to form some remarkably stable zero oxidation state compounds of the lanthanides (see example below).

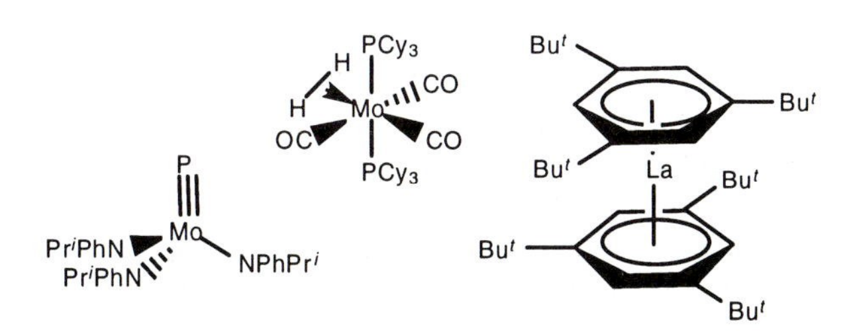

For a fuller discussion see F. G. N. Cloke, *Chem. Soc. Rev.*, 1993, **22**, 17. D. White and N. J. Coville, *Adv. Organometal. Chem.*, 1994, **25**, 36, and T. E. Müller and D. M. P. Mingos, *Trans. Met. Chem.*, 1995, **26**, 1

Z

Zintl isoelectronic relationships

Zintl isoelectronic relationships

Many binary compounds are formed between the Group 1 and 2 elements and Groups 13–15, e.g. Na_3As, $CaIn_2$. The Zintl concept utilizes isoelectronic relationships in order to provide some insight into the solid state structures adopted by such compounds. The electronegativity differences between the atoms are sufficiently large in these compounds that, to a first approximation, an ionic formulation is reasonable, e.g. Na_3As may be written as $3Na^+As^{3-}$. The isoelectronic relationship is then used as a basis for rationalizing the manner in which the Group 13–15 elements aggregate in the solid state. For example, NaSi is formulated as Na^+Si^- and the isoelectronic relationship between Si^- and P suggests that the silicon atoms in NaSi might form Si_4^{4-} tetrahedra analogous to the P_4 tetrahedra in elemental white phosphorus. This geometry is indeed the one observed in the crystal structure. Using similar reasoning, it could be argued that the phosphorus anions in LiP ≡ Li^+P^- are isoelectronic with S and therefore LiP should have P_8^{8-} crowns similar to the S_8 units observed in monoclinic sulfur. Actually, the phosphorus atoms form helical spirals similar to those observed in elemental selenium and tellurium. Therefore, a strict isoelectronic relationship is not observed, but the Zintl concept does suggest possible group structures. A further example is provided by $CaSi_2$ in which the $[Si_n]^{n-}$ component forms a layer structure resembling that observed in elemental arsenic.

Many of the anionic species observed in Zintl phases in the solid state are not observed as stable solution species, because their large negative charges make them very nucleophilic and therefore sensitive to the slightest traces of moisture. Their synthesis in the solid state, which involves mixing and heating under dry conditions the appropriate ratios of the elements, is ideal for creating the anhydrous environment necessary for isolating such reactive species

For a more detailed discussion of Zintl compounds see U. Müller, *Inorganic Structural Chemistry*, John Wiley and Sons, Chichester, 1993

In general for M_xA_y the total number of electrons in the formula unit may be used to estimate the type of ring or polyhedral molecule found in the structure for the A atoms. If it does not form element–element bonds the total valence electron count is 8, but as the number of electrons is reduced from 8 it can be used as an indicator of the number of A–A bonds which are present in the anionic component. Table Z.1 illustrates that when the number of valence electrons per atom is seven, dimeric A–A molecules are formed, when it is six, ring compounds are observed, and when it is five, three-connected polyhedral molecules result. Diamond-like lattices are observed for the Ga^-, In^-, and Tl^- anions in the following: NaTl, LiGa, LiIn, and $BaTl_2$ because of the isoelectronic relationships between them and the Group 14 atoms, Ge, Sn, and Pb.

Table Z.1 Examples of the application of Zintl relationships

Formula	Total number of valence electrons	Structure of anion	Comments
Na_3P	8	P^{3-}	
Na_2S	8	S^{2-}	
$\{Li_2S_2\}$	7	S_2^{2-}	Isostructural with Cl_2
$Fe^{II}S_2$	7	S_2^{2-}	"
LiAs	6	As_4^{4-} ring	Rings or
$\{InP_3\}_{1/3}$	6	P_6^{6-} ring	infinite chains
{CaSi}	6	Zig-zag or helical chains	cf. S and Se
$\{BaSi_2\}$	5	Si_4^{4-} tetrahedra	Isostructural with P_4
$\{CaC_2\}$	5	C_2^{2-} pairs	Isostructural with N_2
$\{CaSi_2\}$	5	Corrugated layers	Isostructural with α-As
NaTl	4	Diamond structure	